I0010964

Make:

ELECTRONIC MUSIC FROM SCRATCH

A Beginner's Guide to Homegrown Audio Gizmos

Foreword by Brian Dewan

BY KIRK PEARSON

FOUNDER OF DOGBOTIC

Make:
ELECTRONIC MUSIC FROM SCRATCH

By Kirk Pearson

Copyright © 2024 by Kirk Pearson.

Foreword copyright © by Brian Dewan.

All rights reserved. No part of this book may be reproduced in any form without written permission from the publisher.

ISBN: 978-1-68045-809-1

Sept 2024: 1st Ed; Oct 2024: 2nd Ed

See www.oreilly.com/catalog/errata.csp?isbn=9781680458091 for release details.

Make: BOOKS
President Dale Dougherty
Creative Director Juliann Brown
Editor Kevin Toyama
Illustrator Maisy Byerly
Copyeditor Mark Nichol
Proofreader Carrie Bradley

Make:, Maker Shed, and Maker Faire are registered trademarks of Make Community, LLC. The Make: Community logo is a trademark of Make Community, LLC. *Make: Electronic Music from Scratch* and related trade dress are trademarks of Make Community, LLC.

Many of the designations used by manufacturers and sellers to distinguish their products are claimed as trademarks. Where those designations appear in this book, and Make Community, LLC was aware of a trademark claim, the designations have been printed in caps or initial caps. While the publisher and the authors have made good faith efforts to ensure that the information and instructions contained in this work are accurate, the publisher and the authors disclaim all responsibility for errors or omissions, including without limitation responsibility for damages resulting from the use of or reliance on this work. Use of the information and instructions contained in this work is at your own risk. If any code samples or other technology this work contains or describes are subject to open source licenses or the intellectual property rights of others, it is your responsibility to ensure that your use thereof complies with such licenses and/or rights.

Make Community, LLC
150 Todd Road, Suite 100
Santa Rosa, California 95407

www.make.co

CONTENTS:

CONTENTS:

CONTENTS:

FOREWORD BY BRIAN DEWAN

Popular Recreational Electronics

If it's not one thing it's another
And there's more'n one way to flip a pancake
Home made, home grown, made by hand
Which hand

You can splice tape with the whole family
In your garage
Helps to have a workbench
Upside down garbage can'll do

If you cob together
Electronic entrails
They'll wanna have a box or something
To put them in

Or a board
Or a few boards or trays
Stacked up
Like a theater organ

A body without a skeleton
Will slip and slide
Exoskeleton or regular skeleton
Will keep you out of trouble

A lunch box full of parts
Oscillators and small loudspeakers
Discreetly tucked away
In scenery for model railroads

Good luck, and have fun
Deploying the stuff you and your friends make
For exciting sonic purposes
Or other purposes

Back issues of *Electronaut* magazine
Loaded with nifty ideas and plans
For people who want to make
Their own TV set or test equipment

Radios, theremins, VCAs
VCOs, white noise, pink noise
Filters, impeders, and preventers
Frequency generating doodads

Unloved microphones
Put to good use once more
A splendid gift
For future generations

No one can stop you
Unless you're too loud
Unnecessarily loud
Enough to draw cops

But then they'll dig
The thing you concocted
Tell you to turn it down
And have a good night

Brian Dewan is a musician, inventor, instrument maker, and performance artist. He and his cousin Leon Dewan perform as the electronic music duo Dewanatron, and their original electronic music instruments have been displayed (and played) at the Armory International Art Fair in New York and Steve Allen Theater in Los Angeles. Their handmade Swarmatron instrument was used by Trent Reznor in the musical score of *The Social Network*. Brian lives in Catskill, New York.

Part I

ELECTRONIC MUSIC FUNDAMENTALS

Y ou can draw the shape of a sound through electricity.

That's a simple but powerful statement! By creating a pattern of electrical flow, we can recreate any sound you've ever heard, or even make a sound that's never existed before. This discovery was met with enthusiasm by engineers and eccentric noise makers with equal vigor. Building electronic instruments is a great way to learn how sound works on a very technical level, a little like what a scientist does. It's also a great way to make really weird sounds that will perplex all your friends, a little like what an artist does.

The first section of this book is dedicated to understanding *what* electronic music is culturally and *how* it works scientifically. And though you don't really have to know what you're doing to make your way through this book, we think it's handy to understand the simple background of the social and scientific legacy you're now a part of. What are you going to make? And how are you going to use it?

The practice you're about to take part in has a rich history filled with characters that—in all seriousness—started out right where you are right now. The greatest innovations in electronic music were often birthed out of accidents—a lot of times by a person that only kinda-sorta knew how to use the equipment. Now, dear reader, it's your turn to join that history.

Which brings me to my first question: When did electronic music *begin*?

01

A PEOPLE'S HISTORY OF ELECTRONIC MUSIC

Nobody knows how old electronic music is.

Or, at least, that's what eleven-year-old me read on a webpage at some point in the mid-aughts. The page, Ishkur's Guide to Electronic Music 2.0, absolutely broke my mind as a child. It was a seemingly endless clickable timeline of electronic music history that jumped from genre to baffling genre. This glorious artifact of the Web 1.0 era eventually went the way of most electronic music: obsolescence. After Flash support was dropped in the mid-2010s, the webpage fell into disrepair, and eventually, linking to it resulted in a 404 error.[1] I will present Ishkur's argument as best as I can remember.

Humans have a great track record when it comes to doing amusing things their survival isn't contingent on. To learn the craft of electronic music is to learn the history of how people misuse existing technology for their own groovy benefits. To study electronic music is to study how people tinker with the world around them—how people reappropriate technologies, sounds, and interfaces to create something of cultural value. The aesthetics of electronic music are ever mutating, which is why it sounds quite different today than it did even thirty years ago. However, the approach and attitude toward electronic music are still the same.

If you're among our younger generations of readers, there's a nonzero chance you might think electronic music came of age in the late 1970s, which isn't the craziest thought. The 1980s were when electronic instruments were inexplicably pushed into the mainstream, suddenly becoming a symbol of popular culture itself. Bands that you might have heard of such as Devo, Kraftwerk, and Parliament somehow captured the popular imagination, turning a previously geeky subculture into something, dare I say, cool?

Programmable drum machines came into existence around then, too, and quickly replaced many a drummer by the decade's end. Another development in the '80s was the birth of MIDI, a fancy system that allows various electronic instruments to communicate with each other, even if they are built by different companies in different parts of the world.

But you, a distinguished and worldly reader, probably know electronic music came long before the '80s. In fact, there's a good chance you've heard of one Robert Moog, a brilliant engineer and synth pioneer who in the late 1960s created the first commercially successful synthesizer. Moog, though not a musician, became good buddies with keyboardist Wendy Carlos, who gave Moog practical suggestions on features that would make his instrument intuitively playable for traditionally educated musicians.

For the first time, you didn't have to have an engineer's mind to learn how to

---

1. A contemporary version of this website—albeit a less fun, less snarky version—exists. The original historical treasure, however, remains lost.

play electronic music—Moog and Carlos brought synthesis out of the lab and onto the pop charts. These instruments became so popular that *Moog* became synonymous with *synthesizer* throughout the early 1970s.

Oh, right, but then there was that curious Russian kid with an aptitude for the cello. Leon Theremin made his mark on the world through an instrument perfect for germaphobes. His 1920 invention, the theremin (what a narcissist), consists of two radio antennae positioned perpendicular to each other. They are proximity sensitive, allowing the performer to play it by waving their hands as if conducting a ghost orchestra. Typically, the left hand controls volume, while the right controls pitch. The theremin was an instant icon, somewhat thanks to its touch-free interface, but chiefly because the sounds it created were so dang *otherworldly* to people of the time. (Even Robert Moog got his start building theremin kits in his garage!)

So, electronic music is about a hundred years old, then. We can't possibly go back further than this, right?

As it turns out, someone else got the jump on Theremin a good quarter-century prior. In 1896, the enterprising Thaddeus Cahill constructed an instrument that even I—a professional ridiculous person—will call truly ridiculous. The Telharmonium was a 200-ton colossus of spinning rods and wheels attached to a musical keyboard. As the keyboardist played, electric motors would spin huge shish kabobs of perforated metal disks. On the perimeter of each disk sat a ring of magnets. If you could spin those disks quickly enough, the changing magnetic fields would move fast enough to produce a musical tone.

Cahill's enterprising plan was to run wires all around New York City, connecting a centrally located Telharmonic Hall (we are not making

The Telharmonium console (hardly pocket-sized)

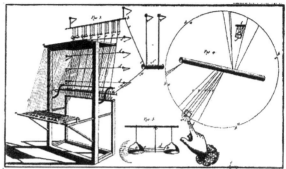

The clavecin électrique (a big hit at parties)

this up) to the houses of New York's wealthiest. The Telharmonium is, in no uncertain terms, an example of streaming media in the late nineteenth century. Unfortunately, the sheer number of wires crammed into such a small space proved too much to handle, as Telharmonium subscribers started to complain about crosstalk with other people's phone calls. Weighed down with the energy costs of operating a 700-watt instrument, the Telharmonium project shut down. No recordings of the Telharmonium were ever made, and the instrument itself was sold for scrap parts in the 1960s.

That's got to be the earliest artifact in electronic music history, right? Well, perhaps you're like my college professor, who claimed that electronic music was invented in 1760 by a priest named Jean-Baptiste Thillaie Delaborde. His gizmo, the *clavecin électrique* ("electric harpsichord"), consisted of several bells on rods that would shake when hooked up to a source of electric charge. Obviously, this wasn't the kind of instrument that would plug into a wall, but it did use static electricity to make a bell ring, so I guess there's something there. Right?

But wait—what about Gottfried Leibniz? In 1665, the brilliant German polymath demonstrated a wild machine called a staffelwalze ("stepped reckoner"): the first calculator that could do all of the four classic mathematical operations you know and love. This machine could play around with numeric data by representing numbers as mechanical components. While Leibniz probably didn't think of his creation as a musical invention, the ability to number-crunch on a machine with reproducible results was the first step in producing digital algorithms to produce ideas we could call "musical."

In fact, to hell with it all—electronic music was probably invented by some Cro-Magnon nerd who was struck by lightning and let out a scream that would have made Little Richard weep.

Electronic music is hard to pin down because it's not a very robust classification. "Electronic music" is a *phenomenally* stupid term. It says absolutely nothing about what the music sounds like, or even how it was created.

Almost all music today is amplified with electric microphones, and processed in ways that you can only do electronically. So isn't *everything you hear* nowadays electronic music?

Even worse, what about the term "music technology"? That's a confusing term because it takes two very hard-to-define concepts and throws them into a catch-all phrase. But it doesn't really tell you anything about what a person does to produce music. In fact, "music technology" mysteriously tends to conjure images of rich people. What comes to your mind? Academics? Corporations? People with knowledge that seems just out of grasp for the rest of us?

While "music technology" does very little to describe what an artist actually does, it succeeds wildly at something else: *excluding scores of people from believing they can participate in it*. The phrase has been weaponized to kick people out of an amorphous, exclusive club, a result that is of particular annoyance to us.

It's ironic that we don't call nylon a "dance technology", although wearing nylon fabric certainly allows dancers to do things they couldn't easily do before. Similarly, we don't tend to think of kids building paper-plate maracas as a technology. Truthfully, in music, there is little distinction between technique and technology. Each is simply a means of exploring a new frontier in your musical journey. We believe that a paper-plate maraca and an expensive 500-speaker diffusion array are equally serious experiments in music.

While we have not arrived at a perfect definition for "music technology," we've come up with one that serves us pretty well: Music technology is the act of finding musical applications for apparently nonmusical things. Banging on an empty tuna can is as much music technology as programming a fancy piece of sampling software, providing you've found a new way to bang on it.

NOBODY TRUSTS ELECTRONIC MUSIC

A unique trait of electronic music—and, especially, the types of instruments we'll be building together to produce it—is that there's no one right way to make it or to build the instruments. The violin hasn't had any significant upgrades in several centuries, yet electronic instruments are *always* changing and being chained together in new ways. Adding a single filter or modulation source to an existing synthesizer can existentially change what that instrument is. Whole genres of music have been birthed into existence due to a small modification of something else.

There is, however, one thing all music technology has in common: people are skeptical of it. Every single innovation in music technology has resulted in the proverbial hurling of tomatoes, dating back to prehistory.

Consider the last time you heard someone say, "A DJ isn't a real musician" or, "All electronic musicians do is push buttons onstage." Next time you hear this, don't curse the darkness. Just tell your confused friend that they are part of a long line of skeptics. The piano, when first invented, was derided as a simplistic button-pushing machine: how could a row of eighty-eight EZ-press buttons possibly compete with the artistic nuance of, say, the flute? More comically, I can promise you that when the first Neanderthal hollowed out a bone and blew into it, there was some schmuck behind them saying, "What a horrible replacement for the human voice," and music innovators have suffered similar criticisms ever since.

Though I dislike the term "music technology," the thing it (imperfectly) describes is always ahead of its time. If your music technology experiments aren't pissing anyone off, you're probably not doing them right. This posits something particularly exciting about electronic music: It's *inherently political*. Every piece of electronic music—just through the act of calling it *music*—makes a statement on what music is and can be. It makes a statement on who (or what) a musician is, and who in that present moment has the privilege of being called a musician.

On that note, what does it mean to be human? Can a machine perform with a human? Is the human just controlling the machine, or is the machine leading the human? What's the difference between an art object and an art-making device? For that matter, what is beauty, and when does it become noise? What does your definition say about your own biases and preferences? Every piece of electronic music poses these queries to an audience, and probably quite a few more. No matter what era you're making music in, none of these questions is trivial.

Creating your own instruments is an amazingly symbolic act. In some unwritten way, you're expanding the definition of music in order to accommodate your own work. Furthermore, you're building an original idea specifically to suit the need you're making it for. Every piece of technology ships free with its own technological code of conduct, which may or may not actually suit the end user's needs.

These days, you're not encouraged to figure out how products work. Devices are designed to be thrown away instead of repaired—often built so they cannot even be taken apart. In recent years, companies have been threatening to file copyright lawsuits against anyone who dares to take their own phone apart, because they could conceivably reverse-engineer it. Consider for a minute just how crazy you'd have to be to try to build something of your own in a world that really encourages you to *please not do that*.

SYNTHESIZERS AS PRODUCTS

Synthesizers aren't much better when it comes to the influence of corporate overlords. Since the early 1970s, companies have put a lot of effort into redefining synths foremost as *products*. Consider the image in your head! A synth is an expensive object made by a team of engineers, put together in an industrial factory, and shipped to wealthy, bespectacled hipsters named Klaus. This is an unfortunate stereotype—one that's been amplified for decades by countless great (and not-so-great) electronics companies. Fortunately, however, a synth does not have to be made in a factory by professionals, and you do not need to rename yourself Klaus to build one.

The truth is that synthesizers are folk instruments played by normal, everyday people and have such a storied history that their inventors are almost always unknown. People tend to forget that for the first eighty years the term *synthesizer* was thrown around, there was practically no commercial market for them. Electronic instruments were seen as novelties—giant, complex toys that were considerably more a scientific marvel than a great-sounding instrument. For well over half a century, a synthesizer was a homemade, raucous, somewhat unpredictable instrument almost always made on a dining room table or

The ARP 2600 is a classic synthesizer, made and sold by a corporation. Not all synthesizers are like that.

in the shed of a budding amateur (such as yourself).

And to this day, synthesizers are *still* folk instruments. Go to any small town virtually anywhere in the world, and I promise you there's some eccentric character soldering capacitors in their spare time. Does this person build synths for convenience? Absolutely not. They make instruments because doing so is *awesome*.

Building your own electronic instrument can be a really empowering experience. If nothing else, it's a fresh reminder that the Korgs and Rolands and Teenage Engineerings of the world, while well budgeted, are hardly the arbiters of new sounds. With surprisingly little up-front learning, you'll have the godlike ability to conjure new sounds out of nowhere—sounds that no preset can make. Better yet, you'll have the knowledge that unless someone else has constructed a circuit exactly like yours, with the exact nuances of the parts you grabbed and stuck together, you'll have the world's only copy of that synthesizer. And that's really cool.

Adobe Stock-Ludovic

THE BANE OF LEARNING ELECTRONICS

If you're anything like me, however, you were probably bitten by the electronic music bug quite some time ago. The only thing that kept me from building my own instruments for several years was that, quite honestly, I didn't know where to begin. A lot of technical documentation is incredibly confusing—even for an expert—and none of it speaks to musicians, only engineers. Furthermore, reading internet forums has proven time and time again to be an absolutely terrible method for learning something technical.

This book hopes to change that.

The laboratory that birthed this book, Dogbotic Labs, was founded in an attempt to teach electronics as a creative art. We believe that building electronic noise machines can be a shockingly liberating practice regardless of your creative background, and that the practice is an exceptional way to convince you that *engineering isn't magic*.

We have written this book for you, dear reader. Given your excellent taste in books, we can only presume you are unreasonably attractive and the life of every party you attend. We have done our best to write a book that appeals to curious, artsy people, not engineers. Our hope is to get you making sounds as quickly and easily as possible. We have filled the book with information that *you* will find relevant—tidbits on how circuits interface with culture, exciting historical sagas, background on particular artworks you can replicate, and a lot of jokes my coworker pleaded with me to leave out. (Sorry, Sean.)

Ever since the majority of synthesizer users started interpreting them as products, corporations have carried out an idiotic crusade to keep them that way. Virtually no commercial synthesizers are designed to be repaired, and a very small number of companies even bother to publish their circuits in the first place. It's well within the interest of the world's corporate powers to make sure our electronics remain *mythically* complicated. The more confused you are about how the built world works, the more inclined you'll be to throw out and replace an item instead of trying to fix it.

Perhaps more dystopically, if we follow this trend to its logical conclusion, it will become easier and easier for folks to ignore how the world is built. What happens when you don't understand how radio transmission works? You get people burning down 5G cell-phone towers, claiming they cause COVID-19.[2] What happens when you don't understand how computer security works? You convince

---

2. Readers of the future, this was actually a really big problem starting in mid-2020.

a huge chunk of the US population that their election votes were not counted, despite absolutely no evidence to support it.

Building a synthesizer, while it may sound a little frivolous, will actually teach you how the world works. Every component in an electronic instrument teaches a fascinating lesson, and for this reason, we've chosen to take you into the technical weeds on every one of our projects.

We're doing this for two reasons:

- **Knowing how your circuits work will make it much easier to troubleshoot them.**
- **It's really, really cool. Honestly.**

If you consider yourself a musical performer and not an engineer (as I did for the first decade of my career), you might find yourself a bit frustrated with electronic instruments. A lot of the time, it can feel like you're at the mercy of the factory foreperson, who limits the sounds you can get out of your machine. But we here at Dogbotic Labs have taught thousands of students how to build synths, and we're continually astonished by how fast people absorb the basic concepts and go on to create something far cooler than anything I've ever done in my career.

A lot of synth books teach you how to build one particular instrument. This book, instead, teaches you how to make several parts and encourages you to connect them in whatever manner suits your own creative wishes. Other books may teach you how to paint a portrait. This book will teach you how to make your own paint.

We also should mention that the circuits in this book exist in the collective consciousness of all who have come before us. The building blocks of these circuits have been in use for well over a half-century, and these circuits were developed through thousands of people experimenting, sharing, and getting excited about other people's work. As is the case with innovators in any folk culture, the "inventors" of some of these circuits are unknown. We're proud to be a part of that tradition, and we hope you relish it just as much as we have. We hope you make these circuits even more interesting and that you inspire the next generation of noisemakers—these creations are every bit as much yours as they are anyone else's.

OUR ROAD MAP

Here is a quick summary of exactly where we hope to take you in this book:

- In "**Musical Electricity for Electrophobes**," we'll correct all the lies you were taught in elementary school, and go through a logical, simple explanation of the electrical concepts you'll need to know. We'll learn how electrical signals can be used to represent a sound by building a working speaker out of household items, how to listen to music through your bones, how audio cables work, and how to use your headphones as a microphone.

- In "**The Hello World Oscillator**," we'll show you how to make the basic building block of a classical synthesizer—an oscillator. We'll show you what a circuit prototype looks like and introduce you to the family of electronic components: resistors, capacitors, integrated circuits, and a few of their estranged cousins.

- In "**Amplifiers, Reverbs, and Talkboxes**," we'll show you how to build a simple and usable power amplifier that runs on batteries so you can take your show on the road. We'll also build two ridiculously fun projects that use an amplifier—a plate reverb and a talkbox.

- In "**Soldering, Enclosures, and UI**," we'll put our oscillator into a permanent home.

- In "**Chaining Oscillators**," we'll expand on our first oscillator by hooking it up to another oscillator. And then we'll hook that up to another oscillator. And then another. We'll show you how to make a circuit that sounds like a cricket,

a synth that's controlled by a candle, and a chip that can be used to generate subharmonics of your input signal.

- In "**Schematics and Mass Transit**," we'll teach you how to decipher those scary-looking charts you see electricians using, and also how to navigate the Rotterdam Metro.

- In "**Filters**," we'll use our knowledge of impedance to build simple circuits that allow only certain parts of the harmonic spectrum through.

- In "**Harmonization**," we'll show you how to teach integrated circuits some rudimentary music theory so you can harmonize with yourself. (How lovely!)

- In "**Modulation**," we'll create some little doodads to dynamically control aspects of our synth's sound. We'll make light-sensitive LFOs (low-frequency oscillators), a triggerable envelope generator with variable decay, and even a stereo ping-pong tremolo effect.

- In "**Sequencers**," we'll make a programmable circuit that can play back a series of voltages to play a melody, modulate a filter, or do anything else you'd like. As a bonus, we'll build a second circuit that can compose interesting-sounding melodies automatically.

- In "**Electronic Percussion,**" we'll build a little drum machine complete with kicks, snares, and cymbals galore.

- In "**Phase-Locked Loops**," we'll take a deep dive into the CD4046 phase-locked-loop chip and show you how to use it to multiply frequencies, slew voltages, and make a sample-and-hold module.

We've also included a "boss" chapter, "**The Dogbotophone MK1**," to show you how to create a massive electronic orchestra complete with multiple voices, drums, and self-patching sequencers that will compose for you on the fly. This project, while large, incorporates aspects from every other chapter—it's a good project if you'd like a challenge after working your way through the rest of the book.

The book ends with a one-two punch: first, a concluding chapter called "Thoughts on Automation," discusses the context of electronic music and capitalism. Finally, appendices cover a lot of helpful resources—including a

discussion on where to get parts, an overview of the integrated circuits we use in the book, and a selection of electronic music we think is worth a listen.

Throughout these pages of theory, projects, and the occasional history lesson, we also included sidebars that highlight a number of our synth-building friends. They'll show you the practical applications of these projects—the kinds of art people are conjuring out of a tabletop full of lifeless wires.

We've written this book to help you get started in building your own electronic instruments, but also to lead you to question the politics of your own creative practice. We hope to bring you into the fold of the larger instrument-inventor community and to give you some historical and social context about the space you now occupy. Deep down, we really truly believe that instrument building is an important, empowering activity, and that the more experimentation there is in the world, the more aware and compassionate we all become.

Finally, if you are reading this introduction in a big-box store and wondering whether you should slip this book under your shirt and walk out with it, we won't tell anyone. ♪

Important Symbols

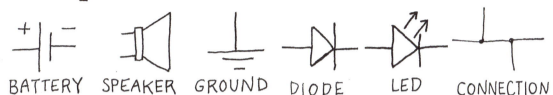

BATTERY SPEAKER GROUND DIODE LED CONNECTION

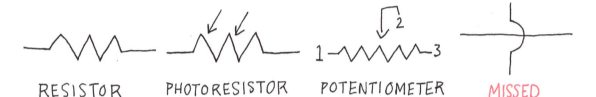

RESISTOR PHOTORESISTOR POTENTIOMETER MISSED CONNECTION

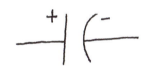

UNPOLARIZED CAPACITOR

POLARIZED CAPACITOR

VERY POLARIZED CAPACITOR

NPN TRANSISTOR

Unit Conversions

NUMBER	NOTATION	PREFIX
1 billion	10^9	"giga-"
1 million	10^6	"mega-"
1 thousand	10^3	"kilo-"
1 thousandth	10^{-3}	"milli-"
1 millionth	10^{-6}	"micro-"
1 billionth	10^{-9}	"nano-"
1 trillionth	10^{-12}	"pico-"

EXAMPLES: There are 1,000 picofarads in 1 nanofarad.
There are 1,000 kiloohms in 1 megaohm.

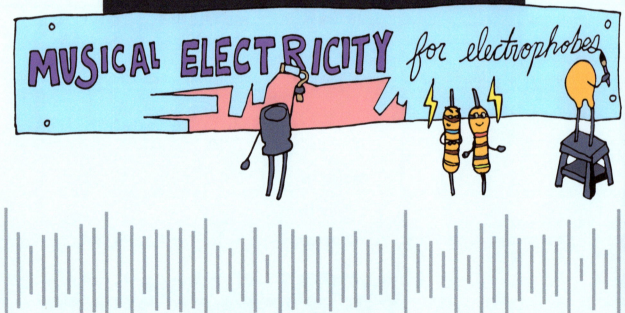

A good two million years ago, one hairy *Homo habilis* smashed a rock against another rock. Things have been very complicated ever since.

Modern humans are incredibly well trained when it comes to mechanical systems—things like rocks smashing into other rocks. As a kid, you learned pretty quickly that objects have mass: the more mass they have, the harder they are to push. You learned to stay away from falling plates in the kitchen. You knew the faster you bicycled, the easier it became to balance. Unfortunately, none of this knowledge is going to help you in your electronics journey, because electricity doesn't work like any of that. At all.

If you share our sense of humor, there's a chance you might have heard of the cartoonist Rube Goldberg at some point. His cartoons depict ridiculously overengineered inventions that accomplish an ultimately trivial task, generally posing many more problems than they solve. What I love most about his machines is, when you view Goldberg's cartoons, how immediately comprehensible most of them are. They're fanciful, sure, but they still operate on basic mechanical principles even children can understand.[1]

Despite what you may think, anything that uses electricity—from hair dryers to microwaves to your computer—works just about as straightforwardly as one of Rube's imaginary inventions. Sure, your computer has a large number of parts, but the stuff that's going on inside is every bit as cartoonish. It's likely you grew up thinking electronics was a miraculous discipline reserved for some higher order of being with an unimaginative T-shirt. Truthfully, electronics isn't more complicated than the mechanical world, per se—it's just operating by a totally different set of rules. We hope that for you, within a few chapters, looking at a simple electrical circuit will be no different than sussing out a Goldberg machine.

Here's the secret: Electricity isn't a thing.

That is, electricity isn't a thing you can see. It's not a thing you can hear, smell, or touch. Instead, electricity is an event. Electricity is simply the thing that happens when electric charges move around.

Unlike mechanical tools, electricity has been known to humanity for only a few thousand years. As far back as classical Greece, we have records of a mysterious, ghostly force we now call "electric charge". The Greeks weren't operating refrigerators or anything, but they were using their cursory knowledge of this force to do some close-up magic.

1. We tried to get one of Rube Goldberg's cartoons in here but it was insanely expensive. Sorry.

The basic party trick goes like this: You stroll on down to the agora and spend some of your hard-earned coin from harvesting your figs on a piece of fur and a chunk of amber.[2] Later, at your nightly orgy, you pull out these exciting new finds and dazzle your guests by vigorously rubbing them together. They watch in amazement as you hold the amber near their neck hairs and the hairs stand up on command. You are thereafter invited to all the dinner parties from then on, dying from smallpox at the advanced age of twenty-six.

Over the years, this trick became so popular that this situation became common:

Kostenlos: Hey, Demitrios, that was a really nifty trick you did there! What do you call it?

Demitrios: Uh . . . I call it the . . . uhhh . . . amber trick.

And, lo and behold, the Greek word for "amber" eventually became the name for this whole category of prestidigitation. That word? *Elektron*.

Moving charges are what we call electricity. But what is "electric charge," anyway? Good question! Unfortunately, after thousands of years of playing with it, we still don't know what it *is* (in a philosophical sense). What we do know is that rubbing some fur on some amber (or, for a more contemporary audience, rubbing a balloon against your sweater) conjures two spooky, mysterious forces. We call them positive and negative charge.

Charge is a property of matter—just like mass, color, or shape. Everything has an electrical charge—it could be positive, it could be negative, or it could be zero—but everything has it. When you rub a balloon on your sweater, the act of rubbing causes the balloon's electric charge to drop from zero to negative. The sweater, meanwhile, changes from a neutral object to something positively charged.

Positively charged things are attracted to things that are negatively charged, and literally tug on them from a distance. This means you can use the balloon to make your hair stand on end, or pull up grains of salt like a tractor beam, or make water coming out of your kitchen tap bend to one side. Meanwhile, two things that have the same charge repel each other. Rub two balloons on your sweater and try to hold them together, and they'll actively push each other apart.

If you've ever played with a balloon or worn a sweater, this shouldn't be that much of a surprise. Although modern physics doesn't have a simple answer to what charge *is*, we can easily explain what charge does. Everything you can

2. Hardened tree sap. See *Jurassic Park* (1993).

touch (amber, balloons, your car, a giant squid) is made out of atoms. Atoms have two crucial parts to them: In their center, atoms have a nucleus made out of even smaller particles called protons and neutrons. Protons have a teeny-tiny positive charge to them, while neutrons exhibit neutral charge.

Because charges of the same kind repel each other, those protons are *fighting like crazy* to leave the nucleus. The only reason the nucleus stays together is because of an incredibly strong force called, uncreatively, "the strong force". The neutrons act as a sort of padding that buffers the protons as they're squeezed in place. Bunch more protons in a nucleus, and not only will you change the identity of the atom, you'll also make the nucleus increasingly positively charged.[3]

Meanwhile, outside the nucleus floats an amorphous cloud filled with particles called electrons. Electrons have a slightly negative charge, just enough to cancel out the positive charge of a proton and make the atom electrically neutral. Electrons are quite interesting because, unlike protons, they can be knocked off their home nucleus without much work. How do you pull an electron off? Easy.

---

3. Side note: The act of cramming more protons in a nucleus (or pulling them apart) is what nuclear physics is all about.

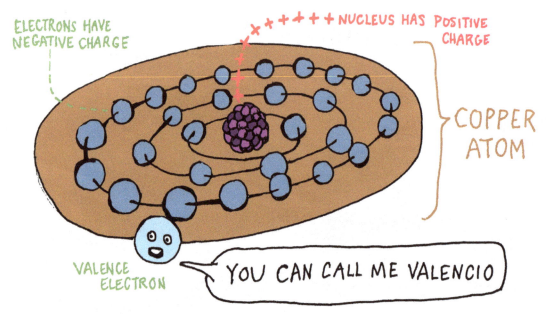

ELECTRONS HAVE NEGATIVE CHARGE

+ + + + + NUCLEUS HAS POSITIVE CHARGE

COPPER ATOM

VALENCE ELECTRON

YOU CAN CALL ME VALENCIO

Rub a balloon on your head.[4]

The act of rubbing a balloon on a fuzzy thing does something amazing: it literally rips electrons off the atoms in your hair and deposits them on the rubbery exterior. Just like that, by removing a couple electrons from your hair, you're now the proud owner of a positively charged coif. Your hair is now teeming with positive ions just aching to make contact with any negative charge in its vicinity. While you might not feel it, your hair is actively tugging on everything around it that has even a slightly negative electric charge.

The balloon attracts your hair because its negative charge is pulling to the positive ions in your hair. The balloon sticks to the wall because it's attracted to the nuclei of the atoms in your wall. Put two rubbed balloons next to each other, and they'll fly apart—due not to differing personality types, but to the fact that they're both negatively charged and, thus, repel each other.

Now you might say to yourself, "Self, isn't it odd how the balloon always becomes negatively charged and my hair becomes positively charged? Why doesn't my hair ever become negatively charged?"

This is an excellent question, and the honest answer is, modern science has no idea. We don't know why the amber steals all the electrons and the fur always loses them. If you do some research into the "triboelectric effect", you'll find a handy-dandy list of materials and which ones are likely to turn positive versus negative. But why this happens is still anyone's guess. Perhaps someday you, brilliant reader, might figure out the answer.

4. If you have hair, that is.

The act of pulling electrons away from one thing and putting them on another thing has a familiar name: static electricity. Static means "sitting still," and refers to the fact that once you rub the balloon on your hair, it remains negatively charged until you give those electrons somewhere to escape to (e.g., you touch it to your friend and give them a shock and irreparably harm your friendship).

You notice static clings on your socks in the dryer; that's because the tumbling causes your socks to exchange electrons—causing some items to be unfashionably stuck to others. You get a static shock when you touch a doorknob after walking around in your wool socks because negative charges build up on you and quickly fly to the doorknob once you get close enough.

The ability of something to gather electric charges and store them for later is called capacitance. We'll talk about how to use this ability for musical purposes in the next chapter.

Despite its appearance in many a kindergarten curriculum, *static* electricity doesn't mean "safe electricity"—the iconic bolt of lightning during a thunderstorm is static electricity, and is obviously not safe for children to play with. In a storm cloud, water particles are violently moshing around and bashing into each other. Just as in the balloon trick, this causes electrons to be popped off some of their mother atoms and recombine. As storminess continues, negative charges begin to gather in the bottom of the cloud, leaving the positive nuclei to "float" to the top.

As the negative charges get stronger and stronger, they start tugging on positively charged air molecules ,leading all the way from the base of the cloud to a neutral point below (e.g., a tree, a church steeple, a tower of terror). The approximately 1 zillion electrons[5] in the cloud move incredibly quickly through this snaky tube of air molecules, causing so much friction that the air molecules instantly vaporize in a bright flash of light. Suddenly, that giant tube of molecules transforms into a vacuum,

5. I'm giving you the latest figures here.

and all the air around it collapses inward, producing—*crack!*—a clap of thunder. Nothing you perceive about the lightning—neither the light nor the sound—is *electricity*. Rather, the flash and boom are byproducts of matter interacting with the flowing charges.

STATIC VERSUS CURRENT

Static electricity, for the most part, isn't terribly usable—you have to be continuously moving something mechanical (such as a balloon, or water molecules) in order to get those electrons to "do work." For all of human history until the past few hundred years, it's all we knew about.

Fortunately, in 1800, the brilliant Italian chemist Alessandro Volta made a discovery: He soaked some cardboard in salt water and sandwiched it between two small plates of zinc and copper. This simple little device creates a chemical imbalance between the zinc and copper, resulting in a difference in electric charge. Volta noticed that if you connected two wires to the two metals, you'd have a self-replenishing source of electrons. This invention is called the voltaic cell, and we still use them today—only now we just call them batteries.

Adobe Stock-vlabo

What can we do with this knowledge? Let's start with a spool of copper wire.

Copper wire is made out of— well, copper, an element that has an atom with twenty-nine protons in its nucleus and twenty-nine electrons flying around in their cloud. Copper is especially nifty because electron number twenty-nine isn't particularly faithful to her mother atom. In fact, it's especially easy to yank a copper electron away from home. If you stick a ton of copper atoms together into a long chain and give those loosely held electrons a push, you can get them to jump from nucleus to nucleus, like a kid on some monkey bars. When electrons move around, they move their negative charges with them. Voilà! Electricity: when electric charges move around.

We call these special kinds of materials that happen to have a loosely held electron "conductors." Good conductors have electrons spry enough to do the monkey bars trick. Bad conductors have atoms that are too tight-knit to bother with such frivolities.

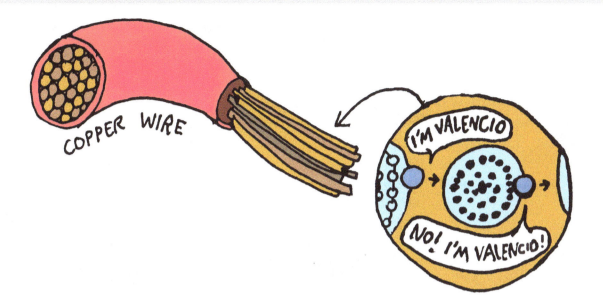

COPPER WIRE

I'M VALENCIO

NO! I'M VALENCIO!

- **Good conductors:** copper, graphite, water with electrolytes (such as salt) in it, gold, Leonard Bernstein
- **Bad conductors:** rubber, wood, silicone, distilled water, James Levine

When you give an electron a nudge in a good conductor, the electrons will continue to hop around as long as you keep nudging. A bad conductor (also called a resistor), on the other hand, will need a much stronger nudge. If you nudge the electrons with more vigor, you'll give those electrons a little more oomph. The worse a conductor a material is, the more you'll have to nudge to get the same result. The scientific name for oomph is—wait for it—voltage.

Voltage is a measurement of potential energy, and we measure it in volts. For instance, if you lift a piano to the top of the Burj Khalifa,[6] you've given that piano a tremendous amount of potential energy—in other words, if you let it go from the top floor, it will hit the ground with a lot of oomph. Dropping the piano from six inches above the ground, however, would obviously result in a lot less oomph. Because voltage by definition represents the difference between charge in two places, you must factor in the two reference points. Falling from the one hundredth floor to the ground is a nightmare. Falling from the one hundredth floor to the ninety-ninth is considerably less bad. In either situation, you're falling from the one hundredth floor. This is pretty analogous to the idea of voltage—if you have one hand on a wire with 100 V, or volts, of potential and the other on a wire with 0 V of potential, you're going to have a tremendous amount of energy pumping through

6. As of this writing, the world's tallest phallic symbol of financial insecurity.

you. If you have one hand on a 100 V lead and another on a 99 V lead, however, the voltage going through you (theoretically) would be barely noticeable.[7]

More literally, voltage is the difference in charge between two places. If we put a slightly positive thing on one end of our copper wire (say, a chunk of authentic Greek amber) and a slightly negative thing on the other end (say, a piece of fur), the spare electrons on one end of the conductor will shoot through the wire toward the positive charges on the other end before the charge between the two points equalizes and the potential energy drops to 0 V.

If we up the ante and attach one end of our copper wire to the base of a strom cloud and the other to the top of a tree, the electrons will shoot through our wire much, much, much, much, much, much, much more vigorously than with our fur demonstration before dropping to zero. We'll most certainly incinerate our wire in the process, create a lot of copper gas, and go into massive debt (that length of copper wire is really expensive). We will also have demonstrated that the difference in charge between the base of a storm cloud and the ground is very large. Thus, there is a much higher voltage between the cloud and ground than there is between the amber and the fur.

The big reason electricity can be dangerous is because it can heat things up very quickly. Your body has far more resistance than copper wire, so it will respond to current by getting warm. If you were to hold a wire in each hand and stick them into the two holes of a wall socket, you'd be in far more danger than if you were to hold both wires in the same hand. In the former situation, you might have current go through one arm and into the other, passing through your heart in the process.[8] In the latter situation, that current will take a short trip through one finger and out the other. This might still hurt, but you'll at least be able to walk away from it.

Tl;dr: Increase the voltage between the two ends of your copper wire, and you'll increase the desire for electrons to flow through it, like chariot racers.

CuRRENT EVENTS

Static electricity, while necessary to understand, is not going to be our primary focus in this book. We'll be making instruments with current electricity, which run off of self-replenishing electron sources called batteries. This is useful if you would like to make an instrument that you can play more than once before rubbing it against something.

The first key difference between static and electric current is that electric

7. Editor's note: This is a thought experiment. Please do not try this.
8. Editor's note: Please, please, please do not try this.

current has to flow in a *circuit*. A circuit, like a circle, is connected on both ends—a loop, if you will. Electric current moves through conductors and always returns to where it started. Break the circuit, and electrons everywhere stop flowing instantaneously.

The gimmick with current electricity is that if we can ensure a constant difference in charge (a.k.a. a constant voltage) on two ends of a conductor, we can make the electrons continuously flow from one end to the other without interruption. Thanks to Volta's battery invention, we can use two chemicals to create a constant potential difference. If we attach each end of our copper wire to the two electrodes, the electrons from one of the chemicals start pushing the electrons toward the other battery chamber. If you replace a 9 V battery with a 90 V source, you'll notice the electrons have ten times the oomph (and your wire might melt or do something fun like that).

Of course, a battery doesn't last forever. Once the two chemicals have equalized their charges, the chemical reaction can no longer continue. The voltage between the two electrodes drops to zero—and we notice because our battery stops working. Some batteries are rechargeable—when you apply current to them, the chemical reaction literally reverses, allowing you to use the battery over and over again.

Current is a measurement of the rate of electron flow. It answers the question, "How many electrons are moving by every second?" We measure current in *amperes*, or *amps*. The more amps, the more electrons are flying by every second. The New York Marathon, with thousands of runners, is a "higher-current" race than the marathon held in Marfa, Texas, with zero runners. While *voltage* refers to how much potential energy sits between two places, *current* pertains to how much charge is moving.

For instance, a high voltage across a good conductor like gold might have a very high current, but the same voltage across a bad conductor like

Birds on a Wire

A riddle for you: Sometimes you see birds perched on high-voltage power lines, miraculously still living. What gives? While the electrical potential between the power line and the ground is quite high, notice none of the birds are touching the ground. Pigeon John can stand on a high-voltage power line because both of his legs are standing on something with the same potential. The 10,000 volts are between the power line and the ground.

If you had two feet on the soil and were to touch John with a long pole, however, those electrons would sail through Pigeon John, the pole, and you, frying you and our feathered friend to a crisp. This is what grounding is—providing a means for electrons to get to a place with a 0 V potential. Our planet, being the place we spend most of our time, is the ultimate voltage reference. Go outside your place of residence and see whether you can find a big metal spike in the ground—that's the grounding for your building, capable of "throwing away" lots of current just in case.

rubber would have a much lower current. Rubber requires more oomph for electrons to exhibit that level of enthusiastic participation. (Current, by the way, is not the same thing as voltage, although in everyday English, we often use the terms interchangeably.)

Thinking about the above analogy, you've probably noticed that—while they are not the same—current, voltage, and resistance are all very intimately related. If you increase the voltage on two ends of a wire, you'll also increase the current going through them. If you increase the resistance of the wire (by replacing it with a worse conductor), the current will decrease.

We measure resistance with a unit called the ohm. The more ohms something has, the better a resistor it is. A one-foot length of rubber tube will have many more ohms between its two ends than an otherwise identical length of copper wire. Ninteenth-century German physicist Georg Ohm wrote up this relationship, and it's called Ohm's law (what a narcissist).

- **voltage (in volts) = current (in amps) x resistance (in ohms)**

If you increase the voltage on a circuit, the current increases proportionally. If you increase the resistance of a circuit between a constant voltage source, the current will decrease.

While electricity isn't a "thing", it is perhaps helpful to use a thinglike analogy here: Imagine a water tank with a hose coming from its bottom. The tank is like a battery, and each drop of water is an electron. The more water we pour into the tank, the higher the water pressure will be at the hose.

As the tank empties, our battery runs out. There's less *oomph* in a run-down battery, so there's simply less water coming out (our current). The water pressure and rate of flow are inherently related!

But if we use a *wider* hose, we can also increase our current. A narrow hose can't fit as much water through it as a wide hose, so the narrow hose exhibits more *resistance*.

- **water = electric charge**
- **water pressure = voltage**
- **water flow = current**
- **hose diameter = resistance**

WORKING WITH MAGNETS

Perhaps we were unfair to the ancient Greeks when we said the amber thing was their signature magic trick. In their defense, they actually had two magic tricks.

Ask Electra Ong

Dear Electra Ong,
Why doesn't the electricity continue to flow out of my wall after I unplug my toaster? Why don't we have to cork it up?
— *Shocked and Confused*

Dear Shocked and Confused,
That's because electricity isn't a "thing"! It's a "when". Electricity is the movement of electric charges (such as electrons). When you unplug your toaster, you're stopping the electron-hopping party dead in its tracks.

Remember: The electrons aren't really being "pumped in" to the toaster, like how water powers a water wheel. The electrons are part of the copper atoms that make up the wires in your wall and in your toaster. By connecting those copper atoms to a voltage source, we cause electricity to occur. Unplugging the toaster doesn't stop the electrons flowing within your wall—they just no longer have any conductive item to flow through. Consequently, we don't need "electrical corks," because air is such a poor conductor already.

Dear Electra Ong,
What is AC? What is DC?
— *Angus Young*

Dear Angus,
AC and *DC* stand for, respectively, "alternating current" and "direct current." So far, we've been talking about direct current, which is when electrons flow in one direction through a circuit. (We call it direct current, because the electrons flow from one electrode on the battery directly to the other. It's a "one-way street" for electrons.) Everything you'll build in this book is DC, as is almost everything powered by a battery.

Alternating current is a *lot* more confusing. In simplest terms, it means that the electrons are not always flowing in the same direction. In an AC system, the voltage on either end of the wire is constantly oscillating back and forth many times per second. At one instant, electrons will be flowing right to left, and in another instant, they'll be flowing left to right. In the Americas, almost all the electricity coming out of your wall sockets is AC, and it's jiggling back and forth sixty times a second. In most of the rest of the world, the AC power oscillates at fifty times a second. In Japan, most confusingly, half the country oscillates at sixty wiggles a second and the other half at fifty.[1]

Why is AC so useful? It can easily be converted to different voltages, and very little energy is lost over long distances. DC power, while easier to think about, cannot be converted without tremendously inefficient power loss. Unfortunately, AC is also harder to engineer and can be a lot more dangerous. Do not play with your AC wall outlet at home.

---

1. This, unsurprisingly, is due to many spheres of influence trying to enact their own systems on Japan at the end of World War II. The result is two very modern electrical grids that are not interoperable due to their different AC frequencies. This occasionally can result in tragedy, such as the Fukushima Daiichi nuclear disaster. While extra power was available, it could not be used because it was created with a different frequency.

Painting by Christoffer Wilhelm Eckersberg

Mr. Hans
Christian
Oersted

Throughout the world, there are deposits of rocks that nowadays we call magnetite. Magnetite is an element that, like copper, demonstrates an interesting property. In this case, the trick is something called paramagnetism.

Many civilizations, including Ancient Greece and China, had access to naturally occurring magnets. As in the case of electric charge, though they didn't know *why* they worked, they quickly figured out what they did. These rocks had two sides to them, and one side would align itself with one side of another rock while repelling itself from the opposite end of the other. It behaved quite similarly to the other gimmick—the electric-charge party trick—but nobody was really sure how or why.

Electrons have a property called spin. Naturally occurring magnets are made of atoms in which all their electrons are spinning in the same direction. This doesn't mean electrons are literally spinning around like little planets, but at first, that's certainly what physicists thought was going on.

Because of that misunderstanding, all the terms we use for them are borrowed from vocabulary we would use for spinning things. This results in magnets having a north and south pole. Like electrical charge, north poles attract south poles, and poles of the same type repel each other. For thousands of years, this was all we knew about magnets. And then, one day, some guy named Hans Christian Oersted had to go and blow the whole thing way into proportion.

The year is 1820; the place, Copenhagen, Denmark. Oersted, an overworked physics instructor, is about to show off a nifty electricity demonstration to his students. On his table sits a proto battery and a big coil of copper wire. He's hoping to attach the wire to the battery and show his students that the voltage induces current to flow in the wire, and that the resistance of the wire will cause it to heat up. (In other words, he's demonstrating the phenomenon that occurs in your toaster.) However, as soon as Mr. Oersted hooks up the wire, he notices something far more interesting. His pocket compass, lying atop his desk, inexplicably, and without warning, twitches. Oersted disconnects the wire, and, there—the compass twitches again. By now, students are getting bored and confused, but Orstead doesn't mind. He's just advanced physics by a good fifty years. Certainly, he'll get tenure now.

You can easily replicate this groundbreaking experiment at home—briefly attach a coil of wire to a battery and hold it next to a compass. Whenever the current starts or stops flowing, you should either see your needle jump (or you should get a new compass[9]). Even more dazzling is what happens when you flip the battery around. Notice that if you send current through the wire backward, the compass still twitches—but does so in the opposite direction. *Fascinating*, Oersted thinks.

What Oersted stumbled upon that day is perhaps one of the most important and seldom-discussed experiments in the history of modern physics. Almost everything that came after that—the nature of space, light, general and special relativity, and basically all modern communication and electronic technology as we know it—starts with that twitching compass demo. *Oh min gud*—said Oersted—could it be that electricity and magnetism are in fact manifestations of the exact same thing?

Turns out, he was right.

The relationship between electricity and magnetism actually boils down to a fairly straightforward concept: Whenever electrical current moves through a thing, that thing becomes a magnet. If you're a physicist, you know this as Faraday's law of induction. Put simply, moving charges induce a magnetic field in the thing they're moving through. This is why the compass twitches:

1. The voltage on either side of the wire causes current to flow through the wire . . .
2. which causes a magnetic field to emerge from the wire . . .
3. and attract any other magnets that happen to be nearby.

If you put a bigger voltage source on either end of the wire, you'll see your compass twitch a little farther. The amount of current flowing through the wire

9. This will not work with your compass app on your phone.

A Magnet you Can Put your Refrigerator On

And now, a discussion about geology. About five miles (give or take a few) below you sits Earth's mantle. The mantle is a terrible vacation spot—its temperature is thousands of degrees, which means everything has the consistency of melty cheese, and the pressure would crush you instantly—a little bit like dining in France. On top of that, it's filled with all sorts of hazardous ions swimming around in a deadly plasticky goop orbiting the planet's outer core. This ion soup is filled with electrically charged particles that are all sloshed around as the planet rotates, like a school of fish trapped in a spherical tank.

These moving charges, thanks to Faraday's law, end up inducting a tremendous magnetic field that emerges from Earth at its poles. Want proof? Take a smaller magnet (say, a compass) and notice that it tends to align itself with the much, much bigger magnet beneath you.

Every once in a while, our nearest stellar neighbor, the Sun, spews a giant loogie of charged particles we call a solar flare. As solar flares fly past Earth, some of those ions are magnetically attracted to the poles. These particles fly into the atmosphere at tremendous velocities, utterly destroying air molecules in the process as they fall from the sky. As the air molecules are ripped up, they give off a little bit of low-frequency visible light. We call this the aurora borealis (or the aurora australis, but that one is photographed less often because it's very cold down there).

Adobe Stock-yusufdemirci and Elizabeth

Pictured: a ton of electric current acting weird with our Earth's internal magnet.

is directly proportional to the strength of the magnetic field. And if you were to sit outside in a field during a lightning storm, you'd notice that each bolt of lightning makes your compass rotate wildly.[10]

---

10. Eh, sure, you can try this one.

EXPERIMENT: THE WORLD'S SIMPLEST SPEAKER

Now that we know how magnets and electricity work in space and beneath Earth's crust, let's put them to good use in a more mundane way.

Before you do this activity, know that the World's Simplest Speaker is not the world's greatest speaker. But it does work—and, better yet, it shows you how straightforward a speaker actually is.

INSTRUCTIONS:

1. Use the sandpaper to scrape the insulation off both ends of the wire so there's a little bit exposed on each end. Usually, the insulation is red and the wire itself is silver.

2. Coil your wire into a two-inch-diameter doughnut, making sure to leave a few inches of wire on either end so the exposed tips have a little bit of slack.

3. Attach a gator clip to each end of the coil (Figure A).

4. Attach the other end of one gator clip to the tip of your aux cable. Attach the other to the sleeve of the aux cable! (Figure B).

5. Time for the magnets. Put a small stack of them on either side of your plastic or paper container (Figure C). They should stick together in place through the surface. Congratulations—you've made a cheapo speaker cone!

MATERIALS:
- A small piece of sandpaper
- A few meters of insulated copper wire (sometimes called "enameled wire" at your hardware store)
- Two alligator clips
- An audio source (a phone or computer with an aux cable will do)
- A few strong permanent magnets
- A plastic container, a paper plate, a piece of cardboard, or anything rigid

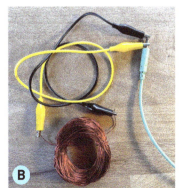

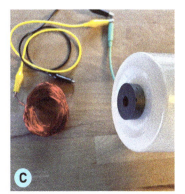

6. Rest your speaker cone on top of the doughnut (Figure **D**).

7. Play some audio through the cable at maximum volume. Hold your ear near your speaker cone—do you hear anything?

D

Why Is This Happening?

Simple! An audio signal is simply a changing voltage. If you send the changing voltage through a coil of wire, the wire will become a magnet with a constantly changing attraction. The higher the voltage sent from your audio source, the more the coil will be pushed or pulled (depending on the direction it's hooked up). By changing a voltage in a musical pattern, we can literally move the world around us in sympathy with electricity. How cool!

Pop Music

Take a look at the back of a slightly better-engineered speaker. You'll notice that—regardless of what speaker you've picked up—there are two tabs on the back of it. These two tabs are attached to little wires called spiders, which in turn connect to either end of a long coil of wire. Every speaker cone in the world—regardless of how expensive it is—works this way.

Attach two gator clips to the tabs on the back of this speaker. Quickly and swiftly, tap the clips on two terminals of a battery and see what happens. Did the speaker pop? Why do you think that happened?

The speaker pops because connecting the battery to the coil of wire creates a current by pushing electrons from copper ion to copper ion. The current, in turn, induces a magnetic field that emanates from the wire. This magnetic field interacts with the permanent magnet inside the center of the speaker, either pushing or pulling the cone up or down. Connect the battery backward and you'll notice the speaker moves opposite to the way it did before (i.e., if it popped "out" before, it should now sink "in").

We have one more step to the puzzle to consider. Sure, a changing voltage can cause a magnet to get "stronger" or "weaker," but how does that translate into the speaker making a sound?

WHAT SOUND IS, ELECTRICALLY SPEAKING

Sound is a pressure wave, and behaves almost exactly like ripples in a pond: Throwing a stone into the water displaces it outward. The more forcefully the stone hits the surface, the higher the waves. Water particles crash into other water particles and continue to push the wave farther away from the epicenter.

A curious aspect to this is that the water molecules hit by the stone are not the same water molecules that are going to splash on the picnickers sitting on the bank of the pond. In a wave, energy—not stuff—moves. Consider doing the wave at a sporting event: The wave moves very fast and very far, but you—a particle—only have to stand up and sit down. Your movement covers a few feet, but the wave moves much faster.

When you clap your hands, you jerk the air molecules around your hands. These air molecules bump into another line of air molecules, which bump into even more, and so on and so forth. Eventually, that wave hits your ear. Perplexingly, a bigger disturbance doesn't mean the sound travels any faster. Throwing a bigger rock in the pond doesn't speed the waves up—it simply makes the waves taller (i.e., it makes the sound louder). The speed of sound is determined by the nature of the material it's traveling through—stiffer materials conduct sound faster. For this reason, sound travels a lot faster through metal than it does through water, and much faster through water than it does through air.

In order to make sound from electricity, we create a voltage wave that is the exact same shape as the pressure wave of the sound. In other words, when air pressure goes

Mono and Stereo

Back in the 1930s, almost all recorded audio was a single-channel signal. We call this a "mono signal" because it involves only one speaker vibrating. But in the 1940s, a brand-new gimmick emerged: music intended to be played out of *two* speakers! Stereo was a big deal because you could create a semblance of artificial space in your recordings by having the sounds move around you.

After stereo became such a big deal, engineers tried to up the channel game for the growing postwar middle class. A three-channel system is typically called an LCR ("left, center, right") and a four-channel system is called a quad. While the three- and four-signal systems didn't last too long in the consumer market, some of their even more intensive cousins did. You might have heard of "5.1 surround sound," a phrase that refers to five audio channels and one subwoofer channel (the "0.1").

Some movie theaters have a 12.1 system, while most top-of-the-line theaters (at least as of this writing in the mid-2020s) go a step further—they employ a system that automatically decodes the sound into as many channels as the theater can supply. Folks who mix for these projects don't even really think in discrete audio channels anymore—they mix sound inside a virtual space, and the speakers "figure out" what they're supposed to do based on where they are. Over the past eighty years, we've gone from watching movies in mono to movies with a hundred-plus channels of sound. Woof.

up, voltage goes up. While the voltage itself is not a sound, it becomes one as soon as it starts vibrating a speaker. When you look at a map of a waveform, you're really looking at a map of how voltage is changing over time. It can also be seen as a map of how your speaker moves across time.

Because sound is just changing air pressure, we can use this simple little system to recreate any sound in the world. This janky collection of metal can reproduce the sound of your child's first words. It can play music that transports you to a version of yourself that you've forgotten about. It can move you to tears, cause you to laugh, or, hell, make you absolutely furious. With just a few elementary electromechanical principles, we've made a device that people have dedicated their lives to. Let that sit with you for a minute.

Sound is really a beautifully simple thing. Every snatch of birdsong, piece of music, declaration of love, or cry of war you've ever heard was merely an act of changing air pressure. Sound is a single line moving through time. Unlike your eyes, nose, and mouth, your ear has to pay attention to only one thing at a time: what's the air like out there?

Ash is a %@¢#$?¢! Robot

In science fiction movies, you often see spacecraft explode in a fiery blast with an accompanying *"boom!"*. Of course, in space, there is no air, nor much of anything, so there's nothing for the sound wave to push through. You can't make ripples in a pond when there's no water in the pond. The gold standard for movie taglines is arguably the one for *Alien*: "In space, no one can hear you scream." (We will ignore the sound that occurs at the end of the movie when Ripley blows up the *Nostromo*.)

EXPERIMENT: MICROPHONE-SPEAKER IDENTITY CRISES

A speaker works by converting a voltage into a location—the voltage tells the speaker exactly where it needs to go at any given instant. But how does an audio cable work?

If you look at the ends of the cables to the right, you'll notice there's a little black circle dividing the metal plug into two parts. That black bit is made of rubber—it's not conductive. (This is important.) The metal bit at the very end is called the tip. The metal bit on the other side of the black circle is called the sleeve. We call this type of cable a tip sleeve, or TS, because it was named by a poet.

Inside a TS cable, you'll find there are not one but *two* wires spanning its length. The tip on one side is connected to a long wire that touches the tip on the other side, and the same thing goes for the sleeve. One of these wires (typically, the tip) carries a changing voltage that serves as our audio signal. The other wire (typically, the sleeve) is a return wire needed to complete the circuit. The reason there are two wires inside a cable is the same reason there are two tabs on the back of your speaker (or, for that matter, two holes in a wall outlet).

They also make smaller TS cables, like the one pictured on the top right which is ⅛ of an inch thick. It still has a tip and sleeve, though, so it works the same as the larger ¼-inch cable. Both of the TS cables shown on the top right work the same way.

Another kind of cable you see often is a TRS cable. Just like a TS cable, it has a tip and sleeve, but it also comes with an additional wire—called the ring, or R. A TRS cable, unlike a TS cable, can send two simultaneous signals along with the return. If you need to send a stereo signal through a wire (i.e., a left and a right channel), a TRS cable might be what the doctor ordered.

These are two TS cables, in both ⅛-inch and ¼-inch flavors.

With a TRS, you can send your left channel through your tip, your right channel through your ring, and your ground (or return) through your sleeve.

TRS cables also come in smaller ⅛-inch packages as well. (These are what we mean when we talk about aux cables.) Remember, despite the difference in diameter, these work identically.[11]

These are two TRS cables, in both 1/8-inch and 1/4-inch flavors—note the "ring" formed between the two black bands on each cable.

Balanced Mono vs. Stereo

When dealing with TRS cables, you'll find that the term "balanced mono" pops up a lot. This means that one of the channels of the TRS cable is receiving an audio signal, and the other channel is receiving the same signal *turned upside down.* What use does this have? Imagine a singer in an arena, hundreds of feet away from the mixing console. While she sings, those hundreds of feet of cable are gunking up her audio—the length of the cable invites pops and fuzz due to physical wear and tear and, because of how long it is, can even pick up stray radio signals.

However, if we use a TRS cable and a preamp, we can send the singer's audio as a *balanced signal.* When it arrives at the mixing console, a circuit flips that secondary signal "right side up." The result? All the pops, fuzz, or stray signals picked up are instantly canceled out. Meanwhile, the singer's voice is doubly reinforced!

To review: You can use a TS cable to send a single, unbalanced mono signal. You can use a TRS cable to send a single, unbalanced mono signal, a balanced mono signal, or an unbalanced stereo signal.

11. Another kind of cable we'd like to mention (although we won't be using it) is called an XLR. This cable looks totally different from a TRS but works the same. Every XLR has three pins—pin 1 is used as the ground, and pins 2 and 3 can send one mono signal each. If you're a musician, you've probably encountered XLRs quite a bit. But as they're cumbersome and not commonly seen on synthesizers, we probably won't mention them again. Sorry, XLRs.

SPEAKER AS MICROPHONE

Now that we know how audio cables work, here's a fancy trick that seems to astonish most people.

INSTRUCTIONS:

1. Clip the two gator clips to the cable— one to the tip, one to the sleeve.

2. Attach the other ends of the alligator wire clips to the two conductive tabs on the back of the speaker (Figure **E**).

3. Plug the cable into the amp with the volume off. Turn it on, and slowly increase the volume.

4. Shout into the speaker. Do you hear yourself?

What's Going On?

Congratulations! You are one microphone richer. Hooking a speaker up backward literally makes a microphone. It's not the world's greatest mic, but it does indeed work. You can 100 percent record a demo tape on a pair of headphones. How nifty!

We've already thoroughly explored the notion that changing voltage can be used to move and shake things mechanically. Attaching a strong voltage signal to the back of our speaker lets us puppet the cone in and out as we see fit. But even more amusing is the fact that we can reverse the process and get voltage from motion. If, instead of pumping current through the wire coil in the speaker, we pushed the speaker up and down, we would still be moving the magnet relative to the coil. Pushing and pulling the speaker cone itself causes the electrons in the copper wire to swish back and forth.

MATERIALS:

- Two alligator clips
- 1/4-inch TS or TRS cable
- A speaker
- An amp

E

- Two alligator clips
- A TS or TRS cable
- A piezo pickup
- Something to beat on (see step 3)
- An Amplifier with a TS or TRS cable input

MICROPHONE AS SPEAKER

A speaker and a microphone—while built for different purposes—exploit the same basic principles. This means we can not only use a speaker as a simple microphone but also use a microphone as a speaker.

WARNING: Do this only with a dynamic microphone—do not attempt it with a condenser or ribbon mic (i.e., fancy equipment). Make sure to keep your volume levels low.

INSTRUCTIONS:

1. Clip the two gator clips onto a TS or TRS cable coming out of a volume-controllable audio source (your phone, your CD player, etc.) just as we did before—one to the tip, one to the sleeve.

2. Turn the volume all the way down on your audio source to start.

3. Attach the other two ends of the gator clips to the tip and ring of the microphone (if it is using a TS or TRS cable), or carefully attach them to pins 1 and 2 (if you are using an XLR cable) (Figure **F**).

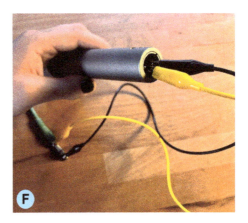

4. Slowly turn the volume up as you play a bit of audio. Put your ear next to the microphone—can you hear anything?

What's Going On?

Dynamic microphones consist of a magnetic slug on springs surrounded by a coiled wire. As you talk into the microphone, the vibrations from your voice cause the magnetic slug to shake around. Moving a magnet so close to wire causes the electrons in the wire to jiggle around in perfect synchrony with the jiggling metal. Here, we can apply a tiny voltage to the coil and make the slug jump up and down. It isn't as loud or clear as a traditional speaker, because the metal slug can't push as much air as a speaker cone. Its job is to be wiggled easily by a voice, not to wiggle easily.

EXPERIMENT: PIEZOMANIA

MATERIALS:
- Two alligator clips
- A TS or TRS cable
- A piezo pickup
- Something to beat on (see step 3)
- An amplifier with a TS or TRS cable input

It turns out there's another way to turn a vibration into an electrical signal, and it's even weirder.

A piezo pickup is a tool of an experimental sound wizard's dreams. You can very easily turn one into a contact microphone that can be used to record very quiet sounds. If you have not heard of one before, you have now.

VIBRATING ELECTRICITY WITH CRYSTALS

INSTRUCTIONS:

1. Clip two gator clips to the cable—one to the tip, one to the sleeve.

2. Attach the two other ends of the gator clips to the wires on the back of the piezo.

3. Tape the piezo to something—a ukulele, a tabletop, a plastic water bottle, a jar of magic beans, you name it.

4. Cautiously turn your amp on and volume up. Excite the piezo by strumming your ukulele, drumming against a tabletop, or shaking your water bottle/jar of magic beans. Do you hear an amplified version of the sound you just produced?

How Does This Work?

A crystal is a solid material whose atoms are built into a highly ordered microscopic structure, creating a very tiny "scaffolding" of sorts. Of course, real scaffolding is held together with pipes and planks of wood and stuff. A crystal is far too tiny for that. The things holding together a crystal aren't things; they're electromagnetic interactions between the ions. These attractive and repellent forces keep each other in perfect balance, ensuring that the crystal lattice doesn't suddenly flop over.

Contrary to the beliefs of your eccentric aunt, crystals are actually useful for more than aligning your life force or whatever. Certain kinds of crystals exhibit a particularly mind-boggling property called *piezoelectricity*, which we can make great musical use of.

Piezos are typically made from quartz—a crystal found all over Earth's surface.

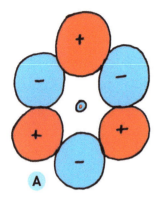

A

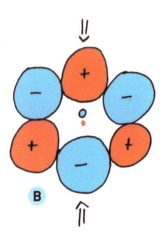

B

If you take a look inside its crystal structure, you'll see a recurring pattern of three positively charged ions (silicon) and three negatively charged ions (oxygen) (Figure **A**).

The positive and negative charges are distributed at equal density. The point in the very center of this shape is the crystal's *center of positive charge*. Stand in the middle, and you'd experience an equal amount of positive electric charge pulling on you from all directions. The same spot is also, unsurprisingly, the crystal's *center of negative charge*. Because of the crystal's apparent symmetry, these points are exactly the same.

Only when this crystal gets squished does something really interesting start to happen: As we press the scaffolding, it starts to collapse inward. The repellent forces between the like charges force the crystal to expand, bulging at the sides (Figure **B**).

Why is this so important? If you slightly deform a piezo crystal, you'll cause its *center of positive charge* and its *center of negative charge* to drift apart from each other. Moving the centers away from one another causes positive charges to accumulate near one side of the crystal, while negative charges gather near the other side. Suddenly, we have a *difference in charge between two places*. Yes, that's right. Squish a piezo, and we have *voltage*.

All a contact mic like a piezo does is provide handy-dandy wires attached to the positive and negative parts of the crystal. Attaching those to an amplifier (like with our gator clips) allows us to move the speaker around whenever any current passes through. If we make our piezo vibrate (by placing it flat on another vibrating thing), we'll induce a current that moves in sympathy with our original vibrating thing.

What Can We Do with It?

As a sound designer, I encounter piezos regularly in my day job. They're super useful for getting rich sounds from otherwise wimpy materials:

- Thwacking a ruler while holding one end against a table sounds only so-so when you listen to it with your ears (or if you record it with a traditional microphone). With a piezo pickup underneath, however, it produces an earth-shakingly deep bass tone—perfect for your forthcoming late-'90s-inspired acid-trance-stomp EP.

- Shaking a carton of orange juice is nothing to write home about, but with a contact mic, it becomes an unmistakable tribute to the high seas. (You can turn the most banal sounds into deliciously strange samples without even trying.)

- An activity that almost everyone loves doing is going on a field trip to the hardware store and constructing a sound box: Find a plank of wood or a box, and collect hooks, nails, springs, doorstops, paint stirrers, wires, pocket combs, tubes, and bobby pins as you please. Arrange them eccentrically, tape or glue a few piezos on the back of the plank or box and you'll have a bizarre percussion instrument that—we're betting—doesn't sound half bad.

- One last piezo trick we must encourage uses Plasti Dip (see "Homemade Microphones" on page 45). This rubber coating is great for piezos, and quickly and easily waterproofs them. Get a few piezos and a can of Plasti Dip, and that's all you need to make a bunch of underwater microphones.

VIBRATING PIEZOS WITH ELECTRICITY

A speaker is a microphone; a microphone is a speaker. A piezo can be a microphone. Does that mean a piezo can be a speaker?

Oh, dear reader, we like you already.
The short answer is yes.
The long answer is *yeeeeeeeessssss.*

Instead of listening to your piezo by attaching it to your amplifier input, try pumping some audio from an aux cable into your piezo. Again an audio signal is changing voltage, which means your crystal will expand and contract as that voltage twists and squeezes the quartz scaffold. It isn't a great speaker, but it can indeed be heard. Only higher frequencies will get through, however; the piezo is relatively lightweight and can't vibrate with the vigor needed for lower frequencies.

Most cheap household appliances with a timer that beeps are using piezos as the beeper. If you have just the right kind of crystal, you can get it to vibrate at a known frequency if you supply it with the right voltage. You also have probably heard of quartz clocks. These keep time with a quartz crystal—in both cases passing the exact right voltage through a piezo allows for startlingly accurate timekeeping.

MATERIALS:

- Two gator clips
- A TS or TRS cable
- A DC motor
- An amplified audio source such as your phone

EXPERIMENT:
SOUND THROUGH YOUR SKULL

Most of the time, when we hear a sound, it's because a pressure wave moved through the air to your ear. But sound doesn't have to travel through air—in fact, you can hear sound in quite a few ways.

INSTRUCTIONS:

1. Attach two gator clips to your audio cable—one to the tip, one to the sleeve.

2. Attach the other ends of the gator clips to the two tabs on the back of your DC motor (Figure **A**).

3. Plug the cable into your audio source and blast some audio through the motor. Do you hear nothing? Great! You're doing it right.

4. Now, bite down on the tip of the motor (Figure **A**). (Yes, we are serious.) Do you hear anything different now?

A

How Does This Work?

Your DC motor is twitching in sympathy with the audio signal. It's not moving enough air for you to hear it with your ears, but when you bite down on it, the twitching sends a pressure wave through your teeth and right into your skull. You are listening to music via bone conduction.

A DC motor is deviously simple. Inside, there are two parts—a rotor and a stator. The rotor is the part that rotates. The stator is the part that stays stationary. Inside the metal chassis of the stator sit two strong permanent magnets. On the rotor sit three small coils of wire. When a voltage source is connected to the rotor, those coils of wire turn into magnets one at a time, driving the motor to spin. The voltage signal coming from your audio source, however, is constantly jerking back and forth. This means the motor isn't so much spinning as it is literally "rocking" along with the music.

What Can We Do with It?

Ever heard of a spring tank? It's a reverberation effect that literally uses a signal to shake a spring. With a DC motor and a piezo, you can make one:

1. Set up your DC motor to twitch when audio is sent through it.

2. Screw two strong hooks into a plank of wood or something, and very *slightly* stretch the spring out between them.

3. Wedge the DC motor's rotor into one end of the spring. Now, when the motor vibrates, the spring will vibrate.

4. Hooke a piezo disk up to an amplifier and wedge it in on the opposite end of the spring from the motor.

5. Blast some audio through the motor while listening to the piezo's output. Do you hear a "washed" version of your original audio signal?

A MOMENT OF CLARITY

As we mentioned in the previous chapter, these days, talking about electronic music is a bit silly, as pretty much all recorded music is electronic music. Every contemporary recording you hear has most *certainly* been manipulated by electronic means, even the ones that don't seem too obvious. Your favorite folk band most certainly uses electronics to compress its dynamics. That "live" orchestra performance was most certainly multiple performances

spliced together to prevent any audience coughing or cell phone ringing. Performances over video are continuously having their sounds manipulated via electronic means—whether that's "de-noising" your background audio or compressing away a sibilant "sss" sound.

It's hard to remember not only that these technologies are relatively new but also that just a few generations before this book was written, they were active subjects of scrutiny. The idea of electronic music was, frankly, very hard for people in the 1930s to imagine.

And then something tiny happened.

At the risk of avoiding exaggeration, let's just say that World War II upset some people.[12] Suddenly, wide-scale destruction beyond anyone's comprehension became instantly possible. Cities were razed, wiped off the map, and minority populations around the world dwindled. No longer did you have to look your attacker in the eye. Now, you could be vaporized from across the world by some jerk sitting in front of a big, scary button. The threat of world annihilation seemed inevitable, mistrust was rampant, and the gnarly teeth of the war industry were suddenly visible for all to see. How positively stressful. Sound familiar? Could we *please* make some synth music?

The fact that electronic music truly took off after World War II happened for very many reasons. First, electronic technology had improved in leaps and bounds throughout the 1930s as a means to build devices that more efficiently and tastefully reduced fellow humans to a pulp.

After the Allies won the war, electronics were hyped in the West as the tool that saved democracy. Bemused children hastily unwrapped dangerous do-it-yourself electronics kits under the Christmas tree. Words like *computer* and *semiconductor* entered the public consciousness, although most people had no idea what they really referred to. *Electronics* was the buzzword of the day, largely planted there by the titans of industry who ostensibly, to the public, had won a war because they were much better at manufacturing components than the guys on the other side.

But second, and perhaps more interesting, electronic music started its slow climb to inevitability for a more poetic reason. After suffering the trauma of war, a need emerged for a type of popular culture of peak technical and aesthetic difference. Electronic music afforded new techniques that gave way to sounds nobody had heard before. It was music so new and exciting, it wouldn't at all remind you of how you used to have a house.

On an even more profound note, the creation of this music also mirrored a more political aspiration. As we explored, the fundamental idea at the basis of

12. Citation needed. (Actually, just take our word for it.)

Musical Innovation: Homemade Microphones

Artist: Alex Taylor

About: Alex Taylor likes looking to unexpected places for new sounds, which kick-started a love affair with homemade microphones. A trip to the swap meet resulted in an army of old telephones, and Alex quickly realized you could wire the handsets to audio jacks for gloriously lo-fi vocals. Of equal interest was the humble piezo pickup, which can be glued into all sorts of freaky things (such as these rubber hands pictured top right). Another technique Alex found success with was coating the piezos in either hot glue or Plasti Dip (a rubber coating you can apply to tool handles so you can grip them better), which forms a watertight seal around the crystal—an easy way to make a functional underwater microphone for nefarious sound-design purposes.

electronic music is that *a voltage wave can be turned into a physical wave*. We can take a microphone, grab a sound made with moving air pressure from "the real world," translate it into a changing voltage, and then *mess with it*.

As it turns out, it's much easier to get electrons to do your bidding than air molecules. We can send this signal through a bunch of electronic components, altering the pattern of electrons, and then—once we're finished—send that signal through a speaker and hear the result. Electronics lets us abstract sound into a totally different domain, mess with it, and abstract it back.

When you build an electronic instrument, you don't really have to concern yourself with issues like air pressure or the elasticity of materials—things that people who invent string instruments have to worry about all the time. As synthesists, we have a lot more freedom when it comes to sounds we can easily make and the ease with which we can make them. Electronic music is a means of transforming the world into something malleable and sculpting new rules for how that world works. With electronic music, you can reject the parts of the world you don't agree with, and substitute them with your own. 🎵

Part II

SMALLER CIRCUITS

Building circuitry is really frustrating. It's finicky, it's hard to troubleshoot, it involves dinky little parts that are tricky to grasp, and it involves spending a lot of time in a room far away from your loved ones. *Woohoo!* Let's get started!

As with all worthwhile hobbies, electronics has a learning curve. If you get stressed, go take a walk and return to your circuit later. While the initial annoyance is, admittedly, hard for a lot of people to get over, the successes are more than worth the frustration. Building a circuit that works—even if it does something simple—is a really empowering experience.

Part II of this book is all about getting your feet wet with some basic electronic prototypes. We'll learn what electronic components do, how to use them in a simple musical setting, and some simple variations. We'll show you how to build a light-sensitive theremin, a talkbox so you can sound like a robot, a pocket-sized amplifier, a working plate reverb, an electronic cricket, and oh, so much more. Along the way, we'll explain the processes of prototyping, soldering, building enclosures, and reading schematics—all necessary steps on the way to building the electronic instrument of your dreams.

But before we get started, it's time to meet your new (extended) family.

03

THE HELLOWORLD OSCILLATOR!

Hey, I have a riddle for you.

Wait—why are you running away?

OK, there's a good chance that throughout your day, you've interacted with many items—incandescent light bulbs, toasters, and hair dryers—that are referred to as *electrical* devices. There are other things—your phone, your radio, and your microwave oven—that are referred to as *electronic* devices. What is the difference between something *electrical* and something *electronic*?

I'm serious—take a minute to consider your answer. It's a little more nuanced than you'd think.

Both electrical and electronic devices use electricity, but *how* they use electricity is philosophically different. An electrical appliance uses electricity as a means of power. An electronic appliance uses electricity as a language.

Simple electrical devices convert electrical energy directly into another form of energy, such as light (in the case of an incandescent light bulb), heat (in the case of a toaster), or motion (as with a hair dryer, which is a combination motor-toaster). You typically can't do very much with an electric device other than switch it on and wait until your toast is burnt. Electric appliances are useful, but they're hardly *smart*. The user is the only one who does any thinking, and their role is limited to "Switch off the toaster when you see smoke."

An electronic appliance, on the other hand, uses signals of voltage as a means to communicate *with itself*. While electronics are also using electricity as a power source, they, for the most part, aren't using a ton of it. Your computer, for example, works by sending incredibly tiny voltage signals around its motherboard. Different parts of your computer, such as your CPU and the pixels in your screen, respond differently to different voltages. By changing the voltage a part receives, you'll change what the computer outputs. Given this complexity, electronic devices can accomplish complicated tasks an electric appliance can only dream of.

In Chapter 2, we discussed *electrical* devices—things like passive microphones, motors, speakers, and piezo pickups. In this chapter, we'll be changing gears to discuss electronic components—things like capacitors, diodes, and integrated circuits.

Each of these devices changes the flow of voltage in its own unique way. For a rough analogy, think about electrical circuits as a bunch of marathon runners moving in a circle. Electronic circuits, however, are much more akin to runners making their way through a ludicrous obstacle course. Every obstacle they come into contact with makes their body do something interesting and unexpected.

The meat and potatoes of this chapter is building the heart of any traditional synth: the oscillator. An oscillator is just a thing that moves back and forth between two extremes, like an oscillating fan or an Ohio state representative. In this case, we're looking to make an *electronic oscillator*. That name is a fancy way of referring to a circuit that outputs a repeating wave of voltage. We can then use this voltage wave to move a speaker up and down at the frequency of our choice.

THE BREADBOARD

In order to prototype a circuit, we'll need to break out the *breadboard*. A breadboard looks like this:

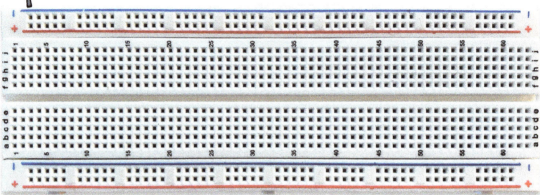

If you have never seen one before, there's a good chance you'll think it looks very intimidating. That's because it does. Despite the scariness, however, you'll be a breadboard expert in five minutes or less.

A breadboard, or *prototyping board*, is kinda like a practice mat—one we'll be using for our obstacle course. We can push components, like resistors and capacitors, inside the breadboard to build a circuit. We can build any circuit we want on a breadboard. If it doesn't work, the breadboard makes it easy to pull out and reposition parts, or even easily replace them if you happen to get a dud component.

Back in the 1940s, prototyping boards didn't exist, so young budding engineers would sneak into the kitchen and steal the family bread-cutting board while their parents were asleep. They could then nail down different wires together on the wood, kinda-sorta using it as a prototyping board like you see today. We don't use literal cutting boards anymore, but we still call them "breadboards"—which shows you that some engineers have an OK sense of humor.

Breadboarding is great for musical experimentation, as it's easy to try out new variations of your circuit before you commit to them. While a breadboard will stay together fine on a desktop, however, they're flimsy enough that they likely will not survive a cross-country plane trip. If you're satisfied with your circuit and want to keep it forever, you can use *solder* (a kind of electrical glue) to give your gizmo a permanent home on a circuit board. We'll show you how to do that in Chapter 4.

Despite the fact that a breadboard has a zillion tiny holes in it, it's actually super, super simple. If we were to rip the top off a breadboard, we'd see something that looks like Figure **A**.

Notice how the holes are connected in rows with metal strips. If two parts are touching the same metal strip, electrons will flow from one to the other.

The two "long rails" have wires that horizontally extend all the way across them.

The many "short rails" have wires that vertically extend the short way.

Normally, the long rails are called *power rails*, and they're where you put the wires on the end of your battery. However, there is nothing inherently different about these rails—they're not made out of a different kind of metal or anything.

We use *jumper wires* to connect different parts of the breadboard together. Jumper wires are just long conductors we can use to "jump" from one part of the breadboard to another.

HOW JUMPER WIRES WORK

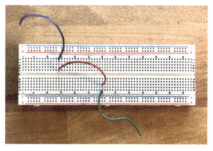

The blue wire connects to the red wire, which connects to the green wire.

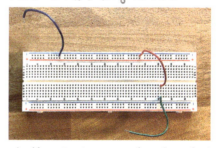

The blue wire is connected to the red wire, which is connected to the green wire.

RESISTORS

When you resist something, you push back against it. This is exactly what a *resistor* does: it pushes back against current.

Resistors are beautifully simple. Down at the resistor factory, they make two delicious stews, one of conductor goop and one of insulator goop. They pick a proportion of these two goops and harden them into resistors. A higher proportion of insulator means the resistor will *resist more current*. This should make sense—imagine passing electric current through 100 percent metal versus a compound that's 10 percent metal and 90 percent rubber. Because electrons flow through metal with more

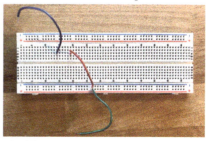

The blue wire is NOT connected to the red wire, which is also NOT connected to the green wire.

ease then they flow through rubber, the resistor with rubber will have a *higher resistance*. Think of resistance as the opposite of conductance.

By changing the ratio between conductors and insulators, you can make different resistors that have different resistances. We measure resistance in ohms (Ω). The more ohms, the more resistance. Sometimes, you might see or hear a resistor referred to with a "k", as in "Hey, buddy, pass me that 10 kΩ resistor, wontcha?" The *k* stands for the prefix *kilo-*, meaning "1,000." A 10 kΩ resistor has 10,000 ohms. Simple.

You can spot a resistor easily because it's decorated with rad retro stripes. These are for more than just decoration, they're actually a code that tells you the value of the resistor. If you want to spend a lot of time learning an esoteric skill, you might be the kind of person interested in learning resistor color codes. Learning which colors mean which number will save you literally pennies over the course of your life and make you totally insufferable at parties. We at Dogbotic Labs are critical of the resistor color-code strategy for a few reasons:

1. The system is comically inaccessible for anyone with vision problems, such as color blindness. And many people with totally fine vision still have incredible trouble discerning the colors.

2. There's no paint-color standard in place. Many companies have red paint that looks like another company's brown paint.

3. Pretty much every other component we use just has a number written on it. I don't know why resistors can't have numbers on them. Is it because printers couldn't print tiny numbers back when they made this system? Beats me.

Why would we want to use a resistor? Many reasons. For starters, a resistor is the most important component in two aforementioned electrical appliances— the incandescent light bulb and the toaster. Both of these appliances exploit the same principle: If a high voltage is placed across a resistor, the resistor won't be able to efficiently pass that energy through it. Most of that energy will be "wasted" as light and heat, the former of which is exactly what we want out of a light bulb and the latter of which is exactly what we want a toaster to do. Another reason to use a resistor, though, is to protect an adjacent component. For example . . .

LEDS AND OTHER DIODES

If you have not heard of LEDs, let me be the first
to welcome you out from under the rock that
has sheltered you for at least the last twenty years.

LED stands for "light-emitting diode." A diode is a component that exhibits very
low resistance in one direction but very high resistance in the other. You can think
about diodes as one-way streets for current, or the spirit component of Harry
Styles. (They allow current to flow in only One Direction.) While diodes are used
for many things, perhaps the easiest to understand them is as valves that prevent
current from "spilling back" into a place it's not supposed to go.

LEDs are a special type of diode that will start glowing when current flows
the right way through them. Notice how an LED has two legs—one long and
one short. The long leg is the positive leg, and the short leg is the negative leg.
I remember this by telling myself, "Positive numbers are taller than negative
numbers."

If you were to spread the legs of your LED ever so slightly and tap them against
a 9 V battery (making sure to put the long leg against the positive (+) electrode
and the short leg against the negative (–) electrode), you'd see the LED light up
really brightly and then suddenly fizzle out. If we want to keep an LED lit up for
more than a split second, we will have to reduce the amount of current flowing

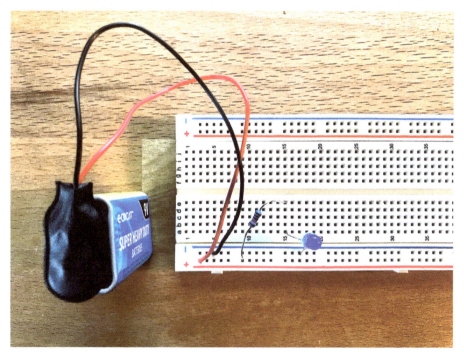

A battery
powering an
LED via a 330
Ω resistor

into the LED. We can accomplish this with a simple resistor.

Remember, a resistor resists current, so not all the current that flows in our wire at 9 V of potential will make it to the LED. Some of the energy is released as heat, which prevents the LED from burning out. (Too much heat can melt the semiconductor interior.) The resistor alleviates some of the stress for the other components.

In the previous chapter, we discussed how all inputs and outputs are interchangeable—a speaker and a microphone can be the same thing, just as a motor and a generator can be the same thing. If we put a voltage across an LED, we can get it to light up. But what happens if we were to shine a light on an LED? What would we make? The answer is, a solar panel.

A solar panel works—in essence—the same way an LED does, but in reverse. On a solar cell, light energy strikes a semiconductor and pushes electrons in the semiconductor up an energy level. In an LED, a voltage knocks electrons in the semiconductor down an energy level, releasing light. If you shine a light on your LED, you will make a voltage between the two legs. Hook up a car battery to a solar panel, and you'll see the panel glow.

POTENTIOMETERS

A potentiometer, or pot, is a resistor you can dial in. Some potentiometers look a little different than others, but they all have the same basic parts. Every potentiometer has a knob on the front that you can spin back and forth, and has three pins—one "middle" and two "sides." These pins—I kid you not—are referred to as the nose and cheeks, respectively. Just like your nose sits between your two cheeks, the middle pin sits between the two side pins. (These are the industry terms—I swear I am not making this up.)

When you spin the knob on a potentiometer, you change the resistance between the nose and the cheeks. Almost every electronic device in your house that has a knob on it is just a potentiometer with a little cap on the tip.

Despite sounding rather complicated, potentiometers are dead simple. If you bust open a potentiometer, you'll find the nose is attached to the knob directly above it, which in turn is connected to what looks like a windshield wiper. This wiper is made out of graphite, which is a swell electrical conductor. On the other end of the wiper is a semicircle of graphite, with each end of the rainbow attached firmly to each cheek.

When you turn the potentiometer really far in one direction, you'll notice that the carbon bridge between one of the cheeks and the nose is really long, while the other is really short. The longer the carbon bridge is, the more resistance the potentiometer will exhibit. The shorter the bridge, the easier it will be for current to flow through with minimal energy loss. This means the more you turn the knob, the harder it is for current to flow from nose to cheek (or cheek to nose!).

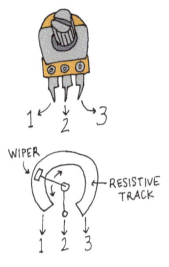

One note before we move on: There are many uses for potentiometers, but for the first part of this book, we'll almost exclusively use them as described as variable resistors. When using a potentiometer as a variable resistor, you need to think about only two of the three pins on the bottom—one cheek and the nose.

Let's say, in the abstract, we want to make an LED-exploder device. (Let me be clear: you don't want to make this device, but it's a helpful thought experiment.)We already know that if we connect an LED to a battery for a little too long, it'll make for a somewhat startling explosion. That is simply because there's too much energy flowing through the LED. But we can keep an LED on with a healthy level of current by adding a fixed-value resistor into the circuit.

But instead, let's connect the long leg of the LED to the nose of the potentiometer. Send the short leg to battery minus, and send battery plus to one of the cheeks on the potentiometer.

Now, consider this—initially, the resistance on the potentiometer will be so high that the LED won't receive any current and thus won't turn on. But as we slowly fade the potentiometer, we'll see the LED glow faintly, then brightly, then dangerously brightly, and then *poof!*—insurance claim.

Potentiometers are one of the most critical components for making musical gizmos, as they provide an easy interface to change how your circuit reacts. Homemade potentiometers are quite easy to make—at the end of this chapter, we'll be creating a janky one to control an oscillator.

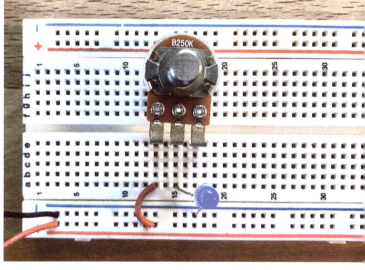

The LED exploder device

CAPACITORS

A capacitor, or cap, is like a bucket. It can be used to store electrons that will be dumped out at a later time. We measure capacitors in farads, named after Michael Faraday. (You might recognize his name from his law of induction, explained in the previous chapter.) The more farads a capacitor has, the more charge it can hold. In other words, more farads means a larger bucket.

What do we need capacitors for? I'll give you three really useful examples:

- Capacitors are used in low-power circuits. Remember that old kitchen appliance you had with a little digital clock on it? I always thought it was funny how you could unplug it from the wall when you go on vacation, but when you plug it back in, it remembers the time perfectly. This isn't magic—it's a really big capacitor that's slowly bleeding charge to power a tiny internal clock.

- Capacitors are used when you need a big dose of electrons really quickly. Have you ever taken a photograph with a disposable flash camera? When you wind the film, you sometimes hear a really high-pitched whooooop sound that increases in pitch. That sound is actually a *flash capacitor* charging. The flash capacitor has a particularly high farad count—when you take a photo, a lethal amount of current zaps through a flash bulb and illuminates your scene with a bang. These capacitors are very dangerous, and we don't recommend taking apart a camera without serious safety precautions.

- Capacitors are used to smooth signals out. Because a capacitor takes a little bit of time to fill (a fraction of a second, in the case of the capacitors we'll be using), it can impose something of a speed limit on the electrons in our circuit. Really sudden moves won't make it through the capacitor, but the general shape of the wave (the slower-moving bits) will make it through just fine. This particular scenario is the secret to how audio filters work. An audio filter can remove certain frequencies from our signal by using capacitors to prevent certain speeds of wiggling.

There are many kinds of capacitors, but you'll get acquainted with only two of them in this book—an electrolytic capacitor, which looks like a black tube with legs, and a ceramic capacitor, which typically looks like a small beige circle with two legs. Both kinds of capacitors work the same way, and in some cases are totally interchangeable. In general, however, ceramic capacitors tend to be smaller than electrolytic capacitors.

Ceramic capacitors are almost never polarized, but electrolytic capacitors always are. This means that they have one long leg and one short leg, which represent positive and negative, respectively. If you plug your capacitor in backward, your circuit will not work. If you're dealing with high voltages and have a particularly low-power cap, your capacitor might even explode! It will be loud and scary and make your room smell like smoke. We do not recommend letting this happen.

INTEGRATED CIRCUITS

The final circuit element we need to talk about in this chapter is also the most complex—the IC, or integrated circuit. The ICs we'll be using look somewhat like this. ⟶

We put this one last because an IC is in fact made up of smaller versions of everything we've talked about already. ICs are little black boxes filled with resistors, capacitors, diodes, and so forth. As the name suggests, they are premade circuits that come in little buglike packages. These days, you can get ICs that are plug-and-play radio receivers, flangers, reverb units, or even fully functional digital computers.

So, if an integrated circuit is just an existing circuit, why do we even have them? Why not just build the circuit in the first place? Do kids these days just not wanna work?

No.

It's because of computer history.

Long ago, before the left and right stereo, computers were humongous electronic machines built out of hundreds of thousands of components— principally, resistors, capacitors, and transistors. While these computers *could* do math much faster than a team of humans, it rarely ever *did*. Why? The dang thing kept breaking.

Engineers realized that even with the best-made components, computers were just too complex to function without a hitch. Even if a super-high-quality capacitor lasts 300 years before expiring, in a machine of 100,000 parts, *that's a capacitor a day you'll have to replace*. And it takes time to find that needle-in-a-haystack capacitor.

One day, however, a guy named Jack Kilby single-handedly[1] came up with a solution. He realized that while a computer at the time might have had a large number of resistors, most of them were a part of a larger circuit that had a dedicated task. "Say," said Jack Kilby, "if we mass-produced single modules of frequently used computer parts, we could make them super easy to replace and

---

1. In the computer-engineering world, this means "with a very large team, including many women whose names are lost to history."

finally be able to check out this new Animal Crossing fad!" And thus the IC was born out of convenience.

Pretty much all the ICs we'll be using in this book were built for old-timey computers. None of them were ever intended to be used as a musical instrument. Despite the fact that ICs come in many flavors, they more or less all look the same. The name of the first integrated circuit we'll be using is the CD4093. (Just rolls off the tongue, doesn't it?)

The CD4093 is called a NAND gate. NAND gates are relatively simple computer components—we'll talk about how they work in Chapter 6. For now, all you have to know is that this was not intended to be used as a musical instrument.

Take a look at this integrated circuit you'll notice there's a little half-circle divot on the left end. Orient this to be on the left-hand side, and make sure the pins are sticking down. This—again, no joke—is called "living-bug position". Flip the IC over on its back and voilà!— "dead-bug position".

With your IC in living-bug position, put your finger on the **lower left-hand corner** and repeat: "This is pin number one." We number the pins on all ICs by starting in the **lower left-hand corner** and counting up counter**clockwise**, as shown in the lower left illustration.

It's absolutely vital you remember this, so take thirty seconds to commit this to memory! It's very easy to get mixed up and presume the upper left-hand corner (or whatever) is where you start counting. Note that this is true for all integrated circuits regardless of how many pins they have. The lower left corner (with the divot to the left) is always pin number 1—whether that's on an IC with fourteen pins or eight.

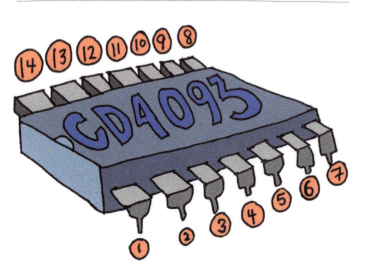

PROJECT: MY-FIRST-SQUARE-WAVE-OSCILLATOR™

This here is the bread(board) and butter of this book. Get this project under your belt, and you'll be *swimming* in project possibilities.

An *oscillator* is a thing that moves back and forth. In geek-speak, an *electronic oscillator* is a conductor that alternates between a lower voltage and a higher voltage. We'll be building a circuit that repeatedly goes back and forth between the extremes of your battery, creating a repeating pattern of 0 V and 9 V. We'll take that signal and send it to a speaker (through an amplifier), making the speaker jump in and out—just like we did in the last chapter. The difference is that with that clicking electronically regulated, we can make our speaker jump in and out hundreds of times a second, producing something we can conservatively call musical pitch. Unless you have incredibly fast speaker-connecting reflexes, you'll probably want to build one of these.

INSTRUCTIONS FOR BUILDING AN OSCILLATOR

Let's get building! Our first foray into synthesis—we'll be making a light turn on and off. It gets more exciting in a few minutes—we promise.

1. Attach your CD4093 to your breadboard, making sure the semicircle is on the left side and the IC straddles the moat. It doesn't matter where on the board you put your IC (Figure **A**).

MATERIALS:

- A 4093 integrated circuit
- A breadboard
- Assorted jumper wires (it helps to have colors that come in pairs)
- One potentiometer (100 kΩ is good)
- Two electrolytic capacitors— one 10–100 microfarads (µf) and the other less than 1 µf
- A 330 Ω resistor
- One LED
- A 9 V battery
- A 9 V battery clip
- Gator clips
- A cable to plug into the amp
- An amplifier of some sort (guitar amps or old computer speakers are great!)

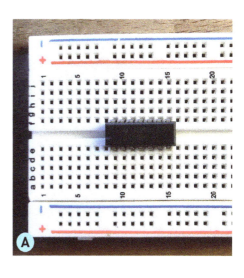

A

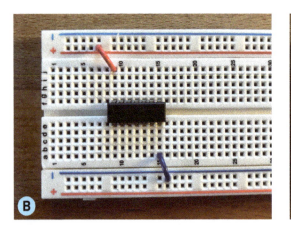

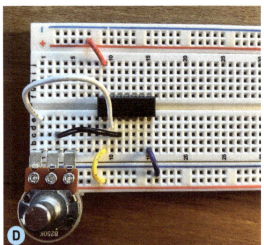

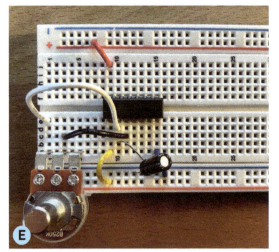

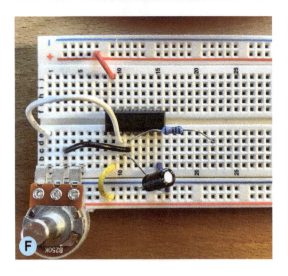

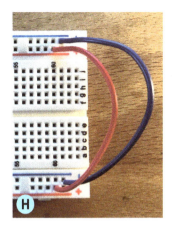

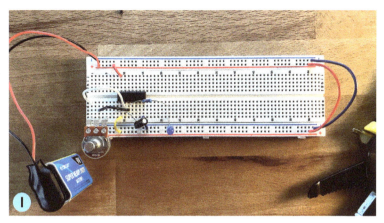

2. Connect pin 7 to battery minus (the black wire) and connect pin 14 (red) to battery plus (Figure **B**).

3. Attach another wire that connects pin 1 to the battery plus row (Figure **C**).

4. Use two wires to attach a potentiometer between pins 2 and 3. Attach one pin to one of the potentiometer's cheeks and another to its nose. It doesn't matter which pin on the IC goes to which part of the potentiometer—just make sure the nose connects somewhere (Figure **D**).

5. Attach the longer leg of a 22 µf capacitor to pin 2 and the shorter leg to battery minus (Figure **E**).

6. Nab a 330 Ω resistor. Have one end touch pin 3 of the IC; the other end can go in any unoccupied row (Figure **F**).

7. Attach an LED to the end of the resistor. The longer leg of the LED should connect to the row occupied by the resistor. The shorter leg should go to the battery minus row (Figure **G**).

8. Bridge the upper and lower power rails (positive to positive, just like before) (Figure **H**).

9. We're ready to test! Plug in a 9 V battery (on a battery clip). The red wire connects to the positive power rail, and the black wire connects to the negative power rail (Figure **I**). Do you see your LED flashing? What happens when you move the knob on the potentiometer?

Oh, God—Oh, No! It's Not Working!

If you've gotten to this stage and you don't see an LED flashing, fear not! Here are the most common problems we've seen:

- Touch your IC. Is it really hot? If so, disconnect the battery immediately! There's a high chance you plugged in the power connections backwards or have a misplaced wire.

- It's possible your LED is flashing too fast for you to tell. To see whether this might be the case, first try turning the potentiometer to see whether one direction noticeably slows down the pulse. If this doesn't work, try replacing the capacitor with one that has a larger value. That capacitor will take longer to "fill up" and thus will blink less frequently.

- Make sure your CD4093 is positioned correctly. Is the semicircle (divot) on the left-hand side? Are all the legs going into their respective breadboard holes? While you're there, did you make sure you indeed put in a CD4093? Different ICs do different things, so they cannot be interchanged.

- Make sure that your power rails are connected from the top to the bottom, your 4093 is correctly attached to the positive and negative voltages in the corners, pin 1 is attached to positive, the capacitor faces the correct direction coming off of pin 2, and your potentiometer connects pins 2 and 3.

- Try a new battery.

- Check the ends of your wires—are they totally free of any insulating material? Sometimes wires can buckle as they get pushed in, and not totally make contact with the metal wiring below.

- Try a new LED, a new IC, and a new capacitor.

- If all this fails, there is a chance that you have a problematic breadboard. It's rare, but it happens. Move everything to a new spot on the breadboard.

INSTRUCTIONS FOR LISTENING TO THE OSCILLATOR

By now, you're probably pretty tired of your strobe light. How can we turn this into something musical?

I'm so glad you asked. It turns out that the same voltage signal that can flash your LED can also be used to make sound. There's one catch—an LED doesn't really need much power to light up, but a speaker needs quite a bit of power to be moved around enough so you can hear it from a distance. In order to hear our oscillator, we'll need an amplifier. For now, we'll be using the same one we used in the Piezomania experiment in the last chapter (page 39). In the next chapter, we'll show you how to build your own.

1. Remove the LED and resistor; we won't need them anymore. Replace the resistor with a jumper wire extending from pin 3 (Figure **J**). (It's the green one in the image.) This wire is the one carrying the oscillating voltage.

2. Clip a gator clip onto the output wire and another gator clip onto a wire going to anywhere on a battery minus row (Figure **K**).

3. Attach each of these wires to the tip and the sleeve of an audio cable—it does not matter which wire goes to which part of the cable (Figure **L**).

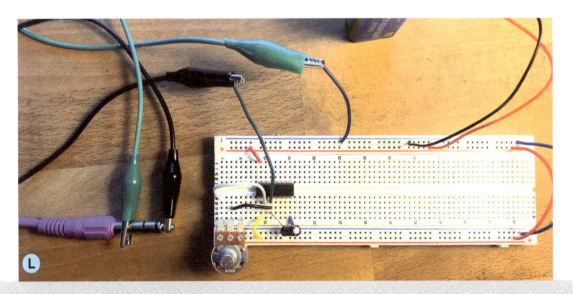

4. Turn on your oscillator and then your amplifier. Do you hear a mysterious clicking sound? This shouldn't really be mysterious—it's the same signal that was turning your LED on and off. But now, instead of watching it, we're hearing it periodically push the speaker cone out of its sitting position.

   This is a pretty mediocre oscillator—how can we get it to make, well, you know, a pitch? The only difference between a slow click and a note you can sing is the speed it's being played back. If we replace our capacitor with a smaller one (say, 0.1 μf), we'll reduce the amount of time needed to "fill" the capacitor with electrons. By making the bucket smaller, we'll need to empty it more frequently! Replace the capacitor and turn on your amplifier (Figure **M**)—does it sound any different?

5. Let's play with the potentiometer for a second. Moving the potentiometer around changes the pitch. Why? Our potentiometer is a variable resistor, which is kinda like a water pipe with a diameter we can change on the fly. If we make the resistance greater, we are making the pipe smaller, which fills up our capacitor bucket more slowly. Both the capacitor and the potentiometer contribute to the pitch you hear in this circuit—by changing either value, you'll change the output.

VARIATIONS ON THIS CIRCUIT

We all know the situation: You build an oscillator, and you immediately try to play Donna Summer's "Hot Stuff" in front of all your loved ones. Unfortunately, the potentiometer doesn't make for a very musical interface—it's just not very natural-feeling to spin a dial around and play melodies. The good news is that there are plenty of things we can replace it with.

VARIATION 1: THE PINCH-O-MATIC

1. Remove your potentiometer from your breadboard—we're going to replace it.

2. In its place, insert two jumper wires. One of the wires should be touching pin 2 on the 4093, and the other should be touching pin 3 (Figure **N**).

3. Turn on your oscillator with the wires disconnected. You should hear silence. Now, pinch the bare ends of both of those wires between your thumb and index finger (Figure **O**). The pitch changes when you squeeze harder!

This happens because *you* are playing the role of potentiometer. The more you squeeze, the more of the wire's surface area makes contact with your skin. More contact means less resistance. For some bonus fun, try licking your fingers and squeezing the wires again. Your saliva, filled with all sorts of tasty ions, is an excellent conductor that decreases the resistance even further.

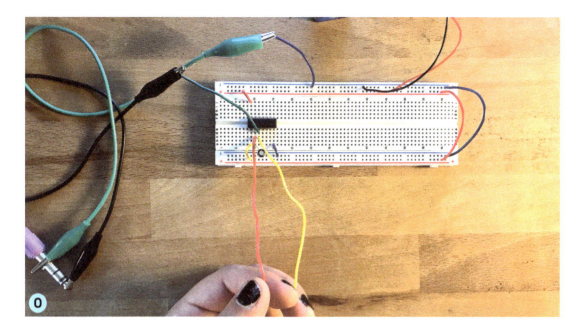

O

VARIATION 2: PLAYING WITH YOUR FOOD

For the health-conscious, here's a dumb party trick. You'll need the following:

- **A breadboarded 4093 oscillator**
- **Two jumper wires**
- **A piece of old fruit**

1. Remove your potentiometer from your breadboard—we're going to replace it.

2. In its place, insert two jumper wires. One of the wires should be touching pin 2 on the 4093, and the other should be touching pin 3.

3. Find an old piece of not-so-appetizing fruit—the older and drier, the better! Push each wire into opposite ends of the fruit so it looks like you're trying to bring it to life.

4. Turn on your oscillator and squeeze the fruit. You'll find that the more you squeeze, the higher your pitch.

5. Try to play your national anthem on your national fruit (Figure **P** on the following page).

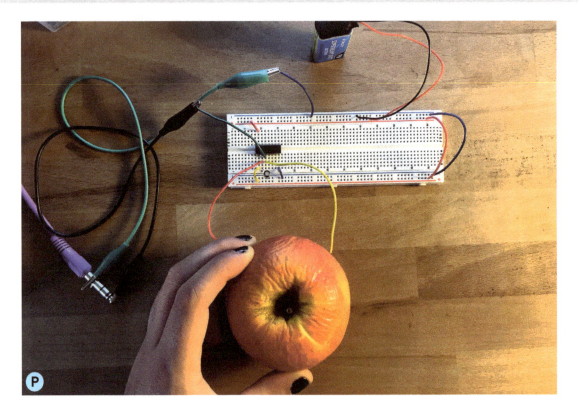

VARIATION 3: THE PENCIL-PUSHER

If you were to open up a potentiometer, you'd find that the two cheeks are attached to each other via a semicircle of graphite that runs around the perimeter. The nose is attached to a single "wiper," also made out of graphite. A potentiometer works by increasing the amount of graphite between the nose and one of the cheeks. The longer the graphite runway, the more resistance between the cheek and nose. Because we can move the wiper around, we can "dial in" our preferred resistance.

Graphite, of course, is also what we put inside pencils—which means we can build our own potentiometer with ease! You'll need the following:

- **A breadboarded 4093 oscillator**
- **A tack hammer and a skinny nail**
- **A wooden graphite pencil**

1. Remove your potentiometer from your breadboard—we're going to replace it.

2. Carefully hammer a nail into the top of the pencil, near the eraser. Make sure the nail goes through the graphite core; otherwise, it will not work!

3. Scribble a big, dark blob near the edge of a piece of paper or cardboard.

4. Attach a gator clip to the big blob and another to the nail in the pencil.

5. Attach wires to these gator clips and send the wires to the spot where our potentiometer goes. One wire should be touching pin 2 on the 4093, and the other wire should be touching pin 3.

6. Turn on your oscillator and start drawing, starting from the dark-blob scribble! The longer the line you draw, your electrons traversing between pins 2 and 3 will meet increasingly greater resistance.

VARIATION 4: A PHOTO THEREMIN

We talked a bit about Leon Theremin's fabulous wiggly instrument, the theremin, back in the introduction. Theremins are played without being touched; the proximity of the performer's hands to the instrument disturbs a pattern of radio waves, which the circuitry can translate to a variable resistance.

What we're about to build isn't *really* a theremin, because pretty much all it does is change pitch based on the amount of light shining on it. While you can play it a little bit like a theremin (by waving your hands around), it's not technically the real deal. But you can counter this argument by claiming that light waves are just really high-frequency radio waves, and who's counting, anyway? Regardless, here's a real crowd-pleaser. You'll need the following:

- **A breadboarded 4093 oscillator**
- **A photoresistor**

1. Remove your potentiometer from your breadboard—we're going to replace it.

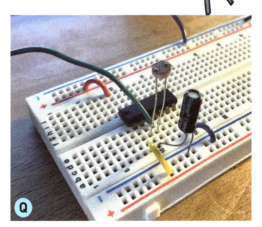

2. Insert a photoresistor, putting one leg into the row that connects to 4093 pin 2 and another leg into the row that connects to 4093 pin 3. Photoresistors are not polarized, so it does not matter which direction it goes in (Figure **Q**).

3. Turn on your oscillator and shine a flashlight on it. See if you hear the pitch change. That's it—you've done it. Pat yourself on the back, you pioneer, you.

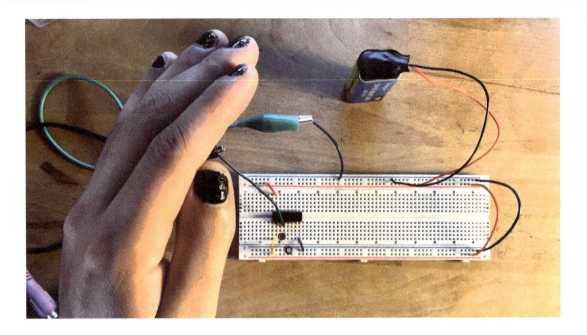

This circuit is no mystery—a photoresistor is another type of variable resistor, just like how we were using our potentiometer. The difference here is that the amount of light shining on the photoresistor causes the resistance to change. Photoresistors are great for fun, arm-wavy control of your instruments, but they're also quite handy for even more complex things such as modulation control. (More on that in a few chapters.)

For now, admire your wonderful little Photo Theremin and marvel at its novelty. But perhaps your circuit leaves you wanting something? Portability? We can totally forgo our big amplifier and create a much smaller one—one that sits on our breadboard and can power a speaker. If that sounds of interest to you, read on. 🎵

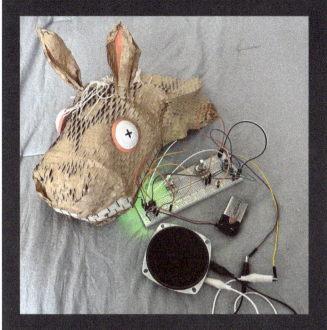
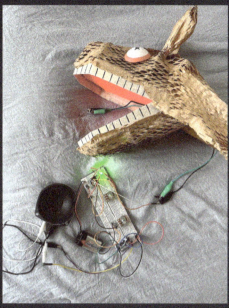

Musical Innovation:
Screaming-Horse Puppet

Artist: Anna Lee

About: Anna Lee is a Los Angeles–based multimedia artist who blends the mediums of animation, illustration, and music to undertake sincere storytelling with lively characters. After she breadboarded a light-controlled theremin (the Photo Theremin variation in this chapter), a jolt of inspiration hit her—chiefly, that if the photoresistor was hidden inside a puppet's mouth, the puppet would "make noise when it was supposed to." Anna returned to class number three with a fully functional screaming-horse hand puppet. We couldn't be prouder.

In the last two chapters, we've been able to hear our creations only by connecting gator clips to a cable attached to a bulky, premanufactured amplifier. If you instead try to connect the gator clips from your breadboard to the tabs on the back of your speaker, however, you'll find you don't hear much of anything at all. What gives?

It turns out there's a simple explanation for this: Our oscillator outputs *very low* current. There are, frankly, not enough electrons whooshing around to do the hefty task of moving a speaker around. By adding an amplifier between the oscillator and the speaker, we can multiply the audio signal by a chosen voltage. This means even more power for the speaker to shake the air around it, which, of course, means a louder volume.

An amplifier is a device that can take a relatively small voltage signal and make it a higher voltage signal. The ratio of the input voltage to the output voltage is called *gain*. A higher gain results in a louder signal. Gain and volume aren't exactly the same, although musicians sometimes use them interchangeably— *gain* refers to the math done by the amplifier, and *volume* pertains to how much of that signal makes it out to the final speaker.

Historically, making an amplifier was hardly an easy task. From the early twentieth century until the mid-1960s, most amplifiers were made using ingenious components called *vacuum tubes.* The tube itself was airtight and made of clear glass, and inside sat two metal electrodes. If you heat up one of the electrodes enough, it will automatically start to exude out electrons, which race toward the other electrode.

Between these two electrodes sits a third piece of metal—our control electrode. If the control electrode receives a positive charge, it will give those electrons a little push, letting them flow through the vacuum tube. However, if the control electrode has a negative or neutral charge, the attraction of those electrons stops—and current grinds to a halt.

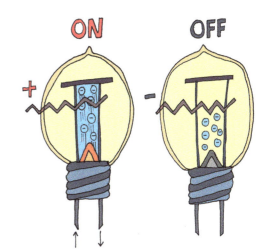

How can we use a vacuum tube as an amplifier? Imagine you've plugged your guitar into a jack attached to the control electrode. As your guitar strings vibrate, they generate a voltage signal that turns on and off hundreds of times a second. If we crank up the gain on the amplifier, the guitar signal will apply the same shape that comes out of your guitar to an increasingly large number of electrons. More electrons means more current, which means louder volume.

Vacuum tubes, while stylish, leave a lot to be desired. For starters, they operate at dangerously high voltages, they're quite expensive, and they're sometimes hard to replace. But even more annoyingly, they don't last too long. At their core, vacuum tubes are basically light bulbs—eventually, they will flicker and die out. All this being said, you'll still find them in fancy amplifiers and effects equipment. Some people just really love the inconvenience, we suppose.

Fortunately, for citizens of the twenty-first century, we present a much better solution: the transistor. Transistors are yet another member of the electronic-component family, and one of the more recent—the earliest transistors were invented in the late 1940s, and most people didn't own a device with a transistor until the 1960s.

How transistors work is a fascinating, albeit slightly complex, process. For now, all you need to know is that their function is quite similar to that of a relay or vacuum tube. A transistor has three legs: a base, an emitter, and a collector. Electrons flow from the emitter to the collector, and you can change how vigorously they flow depending on the voltage being sent to the base.

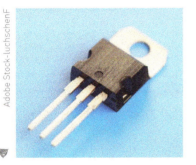

Adobe Stock-luchschenF

Behold the transistor

Transistors were a big deal both because they were a lot smaller than vacuum tubes and because they were much, much more reliable. You might have heard of a transistor radio before—it's a normal radio circuit that uses a transistor (not tube-based) amplifier. While this doesn't sound too fancy these days, this small feature suddenly made radios *super portable* and reliable.

Prior to the transistor, a radio was the size of a microwave oven, took a few minutes for the tubes to warm up, and had to be repaired pretty frequently. By the late 1950s, transistors had turned radios from furniture into pocket-size gizmos. The impact of the transistor on popular music was staggering—it turned music into something portable and even more consumable. Suddenly, new, raucous genres of music came into being now that young consumers weren't listening to the radio in their living rooms. This, and an increase in postwar disposable income, enabled audio technology to explode. The amplifier we'll be making in this chapter uses transistors, as does likely almost every amplifier you've ever used (except for your cousin Eddie's face-melting guitar amp).

Making your own amplifier isn't really a requirement for making a synthesizer. You can always send your signal to an external amplifier. But if you're enthused at the prospect of building an instrument that can make sound on its own, without having to lug around additional gear, this amplifier is for you.

PROJECT: A BREADBOARD POWER AMP

MATERIALS:
- A prebuilt CD4093 oscillator (see page 59)
- An LM386 amplifier IC
- Jumper wires
- Two gator clips
- A speaker
- Assorted capacitors

Our new buddy for this venture is the humble LM386 (Figure **A**).

Like the CD4093, the LM386 is an integrated circuit. Unlike the 4093, it's a lot smaller—a mere eight pins, compared to the 4093's fourteen. Upon this spidery chip sits, believe it or not, an entire amplifier circuit that can goose up an audio signal by 200-fold. The 386 is especially popular for some great reasons:

- **You can get three for a single dollar.**
- **Any place that sells ICs will sell you a 386.**
- **It doesn't burn really hot and give you an owie when you touch it.**
- **It doesn't sound all distorted and gross once you send audio through it.**
- **It requires surprisingly few additional components to run.**

INSTRUCTIONS:

Let's make the simplest possible circuit with the LM386, a mono (one channel of audio) amplifier with a modest gain of 20.

1. Next to our oscillator (sitting on the bottom left of the breadboard), put your LM386 on the breadboard with the semicircle (divot) facing to the left (Figure **B**).

A Our new buddy

B

2. Pin 6 is where the IC gets its power. Tie it to +9 V (Figure **C**).

3. Pins 2 and 4 are ground connections. Tie them to the battery minus row (Figure **D**).

4. Pin 5 is our output pin. Attach pin 5 to a gator clip that goes to the back of a speaker, one that is between 4 ohm and 8 ohm. Attach the other tab on the back of the speaker to ground (Figure **E**).

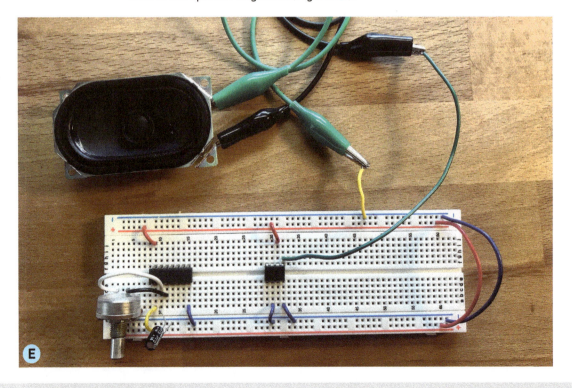

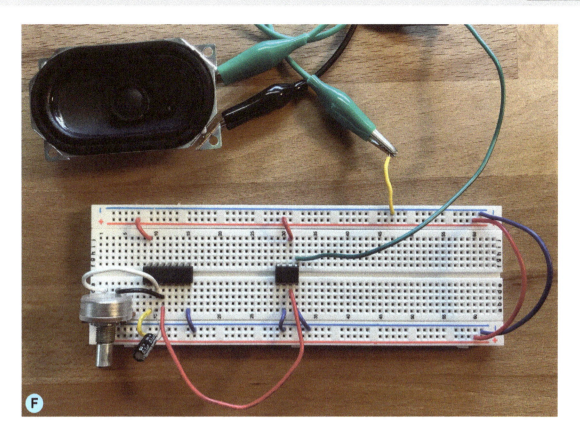

F

5. Pin 3 is our amplifier's input. To amplify our square-wave, we'll send pin 3 of the 4093 into pin 3 of the LM386 (Figure **F**).

6. Attach the battery and take a listen.

Let's say you finish this amplifier and find the results, well, a little lackluster. Using just two capacitors, you can make your amplifier *considerably* louder. Attach a large capacitor between pins 1 and 8 and another between pin 5 and your speaker input (Figure **G**).

G

MATERIALS:

- A prebuilt CD4093 oscillator (see page 59)
- Two LM386 amplifier ICs
- Jumper wires
- A stereo potentiometer
- Two gator clips
- Two speakers

A Stereo potentiometer

PROJECT: STEREO PANNING

You already have a mono amp—why not make it stereo? *Panning* (short for *panorama . . . ing*) is the term for making a sound fly from one speaker to another. When you have two speakers and a panning system, you can make it sound like your oscillator is flying around the front of the room. What marvelous entertainment for your pet Venus flytrap!

The exciting new part we'll become acquainted with is a stereo potentiometer (Figure **A**). Instead of one wiper, stereo pots have two independent resistors that move in lockstep. By twiddling the knob on a stereo potentiometer, you can move two things at once! For this project, we'll be building two LM386 amplifiers and using the stereo pot to change each speaker's gain. By hooking up one of the potentiometers backwards, we can cause an increase in volume in one speaker to correspond to a decrease in volume in the other speaker.

INSTRUCTIONS:

1. We'll start out by building two LM386 amplifiers to the right of our CD4093 oscillator (Figure **B**). Test each amplifier with a speaker to make sure they both work!

2. We won't be sending our oscillator to both LM386 amps, because that might result in unwanted sounds. Instead, we're going to use a NAND gate to *buffer* one of the signals. (NAND is geek parlance for "NOT AND.") Take our oscillator's output (pin 3) and send it to *both* pin 5 and pin 6— the yellow wires in the image (Figure **C**). This effectively duplicates our oscillator, so we have two equally strong square waves coming out of pin 3 and pin 4. Test pin 4 to make sure you hear a copy of your square wave.

3. Now the tricky part: hooking up the stereo potentiometer. The style of stereo pot we're using in these images is intended to straddle the canal

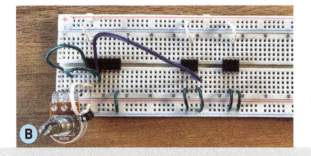

B

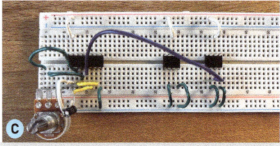

C

D

E

that runs down our breadboard. Three legs (the front potentiometer) sit on one side of the canal, and the other three legs (the back potentiometer) sit on the other side (Figures D and E).

Our left-channel square wave will enter the leftmost cheek at the front of the stereo pot. The nose will connect to the left channel amplifier's input. The rightmost cheek at the front of the stereo pot will connect to ground. The opposite order will follow for the right channel.

You might find it easier to note where the stereo pot's connections are, remove the pot, and then feed your wires in.

4. With both amplifiers ready to go, hook up the two speakers—one for the left amplifier and one for the right. Complete the circuit by connecting the other tabs on the speakers to ground (Figure F).

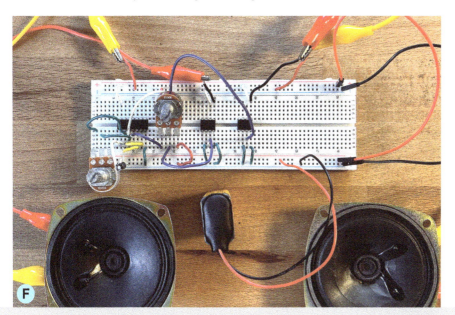

F

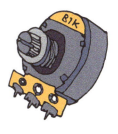

5. Move the stereo potentiometer—you should hear one speaker get louder as the other gets quieter.

ARTIFICIAL REVERB

Shout into a canyon, and you'll hear your voice "smeared" out over several seconds. Shout into a room with carpeted walls, and your voice will be deadened in a fraction of a second. Musicians call this phenomenon *reverberation*, or *reverb*.

We're surrounded by reverb all the time! Every sound somehow interacts with the environment around it, and our ears can interpret these reverb patterns as valuable information. (Consider the difference between the sound of a balloon popping in your bathroom and one popping in St. Peter's Basilica.) For hundreds of years, the only way to produce reverb was to make a sound in a particular place.

Laurens Hammond changed all that.

For a variety of sociopolitical reasons, a huge increase in religious attendance occurred in the early twentieth century, and small churches popped up across the United States. When congregations were unable to fit a traditional pipe organ in these little buildings, inventor Laurens Hammond came up with an idea: What if he could make an organ that produced

The Hammond Organ is an organ without any pipes: rotating magnetic disks create the sounds.

Adobe Stock-michelangeloop

subjectively louder sounds with this newfangled electricity fad?

Hammond's invention—the Hammond organ—does exactly that. Instead of shooting air through a long pipe, the Hammond instead use electricity to spin a set of wheels that had magnets dotted around their perimeters. Each tone wheel would have a coil of wire placed right next to—but not touching—the ring of magnets. The spinning magnetic rings cause electric charges to flow back and forth through the coil. Spin the wheel faster, and it will generate a different pitch. Sensational![1]

Unfortunately, the initial run of Hammond organs wasn't a huge hit for a simple reason: a Hammond didn't offer the same gravitas a pipe organ did. A flimsy tone wheel in a tiny church didn't sound terribly dramatic. This was a horrible PR problem for Jesus. Next, as families purchased their first Hammond organs, complaints began to fly. Customers claimed their organs must be broken, as they weren't hearing the room-filling sounds that were—at that time—intrinsic to the definition of an organ. Add carpeting, low ceilings, and draped windows, and the sound was even worse.

Hammond was at an impasse. Either these unhappy customers would have to rent a warehouse to get the reverberation they wanted, or Hammond would have to develop a system to make an artificial space. Fortunately, he found that Bell Labs in New Jersey had created a nifty device used to test delays on phone calls over long wires. The system used a series of springs to acoustically transmit a signal. Because springs are able to deform and store energy, they can effectively smear a sound that's passed through it. After fiddling with this design, Hammond found incredible success by fitting his organs with a thin metal plate that would vibrate around thunderously. This contraption was named, uncreatively, a "plate reverb", and is still in use in studios worldwide to this day.

Plate reverbs replicate a space that does not exist in the natural world, yet it has now become a characteristic sound we hear in music (earnestly) every day. The important moral of the story is that the presence of reverb—even unnatural reverb—drastically changed how people "read" the sounds of the Hammond organ. Indeed, reverb can totally change how people read pretty much anything . . .

---

1. If this sounds similar to Thaddeus Cahill's Telharmonium instrument mentioned in Chapter 1, you'd be correct. The idea of using a spinning wheel to encode a wave is precisely the same.

MATERIALS:

- A working LM386 amplifier circuit
- An acoustic transducer
- A piezo
- Two gator clips
- A big metal pan, like a baking sheet

PROJECT: A PLATE REVERB

Once you've got an amplifier, it's quite straightforward to start making your own artificial spaces. Our tool of choice is something called an *electromagnetic transducer*, and it looks something like (Figure **A**).

A transducer converts one kind of energy into another—in this case, it transforms an electrical signal into a mechanical one. It's a bit like a speaker without the cone, designed to shake whatever you stick it onto. A plate reverb works by amplifying a signal, transducing the signal into a large piece of metal, and then recapturing the vibration of the metal through a piezo. I'm aware this sounds really dumb, but I swear this is actually how it works.

INSTRUCTIONS:

1. Build an amplifier circuit of your chosen gain. Because we can amplify any signal, this plate reverb will work just as well on your DIY oscillator as it will on your electric guitar, phone jack, or anything else with an analog signal output.

2. Instead of sending the amplifier output directly to a speaker, however, we're going to send it to a transducer. Use the tabs on the back of your transducer to accept the wires coming from your amplifier (Figure **B**).

3. Test your system—play some audio through the amplifier. Does it make your transducer buzz? A note: Your transducer, when not hooked up to a surface, will sound surprisingly quiet. Despite the fact that the transducer is the moving part, it has comparatively little surface area. This means it doesn't actually disturb too much air, which means it's not terribly loud to our ears.

4. When you're ready, adhere the transducer onto an interesting surface. This surface is the *reverberator*, or the thing that we'll be shaking (Figure **C**). Big pieces of metal (baking sheets, wheelbarrows, auto bodies) will give you a classic reverb sound, but plenty of other materials will also sound interesting. Styrofoam will produce a very different-sounding effect, as will a shallow metal dish filled with an inch of water. Some materials— if they have loose objects on top or attached to them—might produce a cool "distortion" effect.

A

B

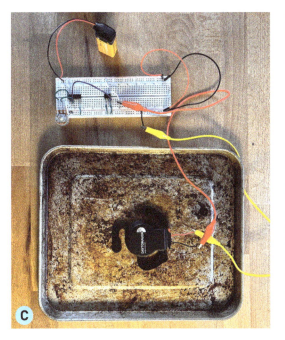

5. Now that we've amplified our signal and used it to vibrate something, our remaining step is to recapture the vibrations from our reverberator. Grab a piezo, plug it in, and either send its output to your "professionally made" amplifier (Figure **D**), or build another one with an LM386.

How Does This Work?

You'll notice that you're in possession of two sound sources—your "dry" sound source (the signal that's going into the transducer) and your "wet" sound source (the signal going into the piezo). Your dry signal is identical to what you're feeding into the reverb unit, and your wet signal is the totally saturated reverb output. By mixing the two layers, you can change the density of your artificial space. A mostly dry output will be a subtle effect, while a totally wet output will be totally spaced out.

Another fun thing to try with your plate reverb is some tasteful *feedback*— taking something that comes out of a machine and putting it back into the input. Just as when you make a photocopy of a photocopy, every iteration of feedback will make your subject appear less and less like your subject and more and more like the nuances of the machine that reproduced it. And though a photocopy will look only fuzzier over time, feedback in an audio system tends to quickly highlight the resonant frequencies of the circuit (resulting in a horrible high-pitched *wheeee!* sound).

Plug your piezo output into your amplifier input and give your reverberator a tap. Your sound system should start wailing like a banshee. Play around with the respective volume of both amplifiers—by changing the gain ratio between them, you can give the whining feedback a bit more dynamism.

THE GLORIOUS INVENTION OF THE TALKBOX

A talkbox is a glorious, ridiculous invention. It's an effects unit that allows you to shape the formant qualities of a sound with your voice, effectively letting your oscillator "talk." The machine takes the form of a rubber or plastic tube that goes in the ~~patient's~~ musician's mouth. It looks really dumb and makes you sound like a robot—and if that's not a good pitch, I don't know what you're doing with this book.

The talkbox we'll be making in this chapter works the same way as Stevie Wonder's or Peter Frampton's—it's just a few hundred dollars cheaper and works at least half as well.

A talkbox is theoretically very simple, and only the slightest bit an optical illusion. It's easy to see Stevie Wonder singing with a tube in his mouth and presume that the tube is a microphone that is sending his voice into some amazing robotization machine. In fact, the tube isn't an input at all—it's an *output*.

The easiest way to wrap your head around how a talkbox works is to blast some audio from your phone speaker, put your mouth up next to it, and mouth out an *"ooh-ahh"* shape repeatedly. You'll very clearly hear the tone of the sound change—the closed *"ooh"* vocal shape reinforces the hollow mids of your signal, while the open *"ahh"* shape brightens it up substantially. The important thing to remember here is that your "mouth-affected" sounds are occurring in the real world, not in the electronic audio signal. To record Stevie Wonder's robot voice, we have to put a microphone in front of him. The talkbox is not where the sound "goes"—it's where the original sound "comes from." Got it? Got it.

The world of vocal manipulation is a wonderful one filled with all sorts of neato pieces of gear. The talkbox is perhaps one of the most basic, although your ear might not know it. Talkboxes are very commonly confused with other fascinating technologies, such as the vocoder and Autotune, although they work in considerably more complicated ways we hope to cover in a future book. This is just a friendly PSA that your talkbox is *not* that.

Stevie Wonder Talkboxin'

PROJECT: A TALKBOX

INSTRUCTIONS:

1. Solder a few inches of wire to each of the leads on your round speaker (Figure **A**). (Don't know how to solder? Check out the next chapter!)

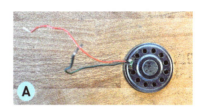

2. Because a talkbox is an *output* and not an *input*, the only part of our amplifier we'll be modifying is the speaker. Here, I've set up a tried-and-true 4093 oscillator going directly into an amplifier, and tested the speaker to make sure it works before removing it entirely (Figure **B**).

3. Preheat the hot glue. Nestle the speaker face down in the funnel so it's straight and snugly seated, and put the whole assembly, nozzle down, into the cup to act as a holder and keep it balanced. Carefully pump a thin bead of hot glue around the border where the speaker meets the funnel to firmly affix the speaker into place (Figures **C** and **D**).

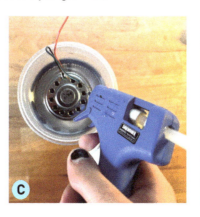

MATERIALS:
- Soldering iron and solder, to solder the wire onto the speaker
- Some stranded wire
- A round speaker with a 2"–3" diameter
- A working LM386 amplifier circuit
- A plastic funnel
- A sturdy cup, to hold the funnel
- Hot glue gun and glue stick
- A small bit of polyester polyfill or a handful of cotton balls
- Silicone caulk/sealant (the kind you get in a tube at the hardware store)
- A few feet of plastic tubing that fits snugly around the nozzle of your funnel

4. Pull the wires to the side and add a good layer of polyfill or cotton balls—something nice and springy. Use some hot glue to get that layer stuck down (as well as you can) (Figure **E**).

5. Extrude a good two inches of silicone sealant over the layer of cotton (Figures **F** and **G**). It will be gross and awful, or else you're not doing it right. Let this abomination dry for at least twelve hours, or risk a sticky surprise:

6. When it's completely dry, take the final assembly out of the cup and invert it nozzle-up onto your work surface. Slide the end of the plastic tubing to the end of the nozzle (Figure **H**). Put the other end in your mouth. Play some audio through the amp circuit while lip-synching the following phrase: **"Silence, Earthling! My name is Darth Vader. I am an extraterrestrial from the planet Vulcan!"**

How Does This Work?

You've probably heard of a body part called the diaphragm. It's a big, dome-shaped muscle in your chest that can pop in and out, pulling or pushing air from your lungs. When we force air out of our diaphragm, we send a current of air into another chamber called the larynx. Once pressure is high enough in the larynx, two folds—our vocal cords—open up like a pair of curtains and oscillate. In our talkbox circuit, the diaphragm and the larynx are modeled by our breadboard. Once the electrons are oscillating, we can send the signal to our talkbox speaker, which vibrates much like our vocal cords.

While our vocal cords help determine the pitch we're singing, they're not providing any of the nuances language needs. Everything else that differentiates speech from a naked oscillation is done by your tongue, lips, and teeth. When you change the shape of your mouth as you talk, you change the tone color of the sound from your vocal cords.

Put your mouth into the shape it makes when you say *weigh* versus *woe*. In *weigh*, your mouth is open wide, allowing some of the higher frequencies to be heard. In *woe*, your closed mouth dulls out those high frequencies. Your mouth is behaving like a *filter* (something we'll discuss in detail in a few chapters). By changing the shape of your mouth, you change which frequencies are reinforced. The talkbox is an artificial larynx, not an artificial mouth. It will generate pitches we can use to "talk," but it will not do the talking for us.

Vocal manipulation is a subject of endless fascination to me, in large part because it's a really ripe symbol. Take a second to think of your favorite song in which the singer doesn't sound like a human. What are they singing about? Does the song in any way highlight their humanity, or lack thereof? In what ways does the character that sings the song embody a machine?

In most places in the modern world, calling someone a robot can either be a great insult or a great compliment. Consider—if your boss says you're a robot at your job, it either means you're incredibly efficient and reliable, or that you're a brainless drone incapable of creativity.

With a talkbox, your voice suddenly takes on a new character that's inherently morally ambiguous. A lot of the time, songs that employ vocal manipulation either take the road of *celebrating* technology (Cher's "Believe," Kraftwerk's "The Robots," the Beastie Boys' "Intergalactic") or wax poetic on how technology might make humans more disconnected (Air's "Run," Newcleus's "Computer Age," Daft Punk's "Digital Love"). Fascinatingly, the same tool works just as well for both arguments.

Despite all this, I believe the most incredible aspect of the talkbox is one that's oddly personal: Though all our vocal systems work more or less the same,

there's a lot of personal identity tied up with one's vocal cords. People make all sorts of knee-jerk assumptions about who *you* are based on your vocal cords. Your age, your race, your gender, and even your sexual orientation have codified vocal types. If your identity doesn't match up with the vocal type society reads you as, simple things like talking on the phone can range from annoying at best to debilitating at worst.

A talkbox changes that.

By replacing a body part that's so tied up with individual identity, you can abstract yourself away from your own anatomy. You can remove a major signifier of your socially determined archetype! Though a lot of people see a talkbox as a device that strips you of your humanity, I believe a talkbox is an amazing tool that can empower you to assume whatever identity you wish. Thanks to the talkbox, you can make the body you've always wanted. 🎵

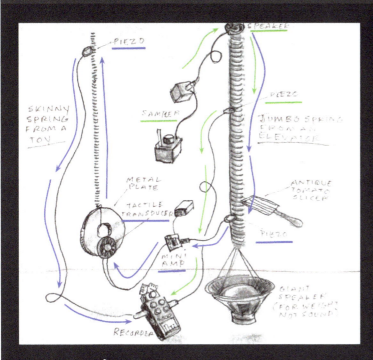

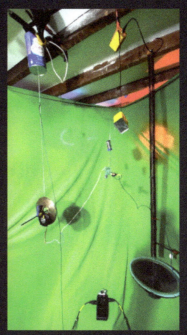

Musical Innovation: Twin Springs

Artist: John Roach

About: Multimedia artist John Roach is fascinated by the interaction of sound and space, and lives to explore their interaction. At least that's what he told his family when he took over the basement to build his final project in our "Ear Re-Training" workshop. John's room-size installation, "Twin Springs", is a reverb unit that uses multiple springs of different gauges in a game of cat-and-mouse. Tones captured through a piezo are played through a speaker, vibrating a spring, which is then itself vibrating a second piezo, which is then amplified and transduced onto a metal plate, which vibrates another piezo . . .

With just a few steps of feedback between input and output, "Twin Springs" can quickly turn *any* slight disturbance into a creaking, groaning, oddly beautiful piece of ambience.

05

SOLDERING ENCLOSURES AND UI

Has this happened to you?

You spend hours on your breadboard, fine-tuning your oscillator until it rings nothing but the truth—a tone so positively perfect and tantalizingly tintinnabulous it is *sure* to bring about world peace. But as you cram it into an envelope to ship to the Kremlin, a bunch of the wires fall out. You realize pretty quickly that a breadboard and jumper wires, while convenient for prototyping, are hardly a stable piece of hardware.

Fortunately for you, there's solder.

Solder is an alloy, which means it's a metal made up of other kinds of metal. You can buy a big spool of it for real cheap, and it looks and feels kinda like copper wire. Solder melts at a relatively low temperature (say, a few hundred degrees Fahrenheit) and congeals back into a solid within seconds at room temperature. It's also very electrically conductive. We can melt solder using a tool called a *soldering iron* or *soldering pencil*.

Electricians can use solder to attach components to other components, typically on a big, flat sheet we call a *circuit board*. Once all the electronic components are attached correctly, your circuit board works pretty much like your breadboard—except, of course, none of the components will be easily removable. This is an object that *can* be packed up in a box and shipped to your favorite (or least-favorite) head of state.

Learning to solder is a bit like learning to juggle. It isn't exactly hard, but it's also something that takes time and practice before you can do it really well. Most people are well on their way to soldering within twenty minutes of practice, and it's a skill that will last you a lifetime. Plus, with some basic soldering skills, you'll be surprised by how many things around the studio you can fix. Old pairs of headphones with a broken speaker are often easy fixes, as are broken audio cables.

While soldering is a basic skill any electrician knows how to do, you absolutely don't need to learn how to solder to make musical circuitry. In fact, we view making sounds and building a finished instrument as totally separate acts. Every activity in this book can be done with a breadboard without ever touching solder. However, our editor heavily encouraged us to put this chapter in, so here we are.

LAB EQUIPMENT

In order to solder, we'll need to go out to the hardware store and pick up the following:

A soldering iron and holster (sometimes called a soldering pencil)

You can get really cheap soldering irons (Figure A) at your hardware store for less than $10—they'll work, but will probably be frustrating. If you can, see if you can get a slightly better iron for somewhere in the $30–$60 range. These soldering irons will have a variety of tips, will change temperature automatically, and may have other wham-o features. We recommend a 20- to 60-watt iron with as pointy a tip as possible.

Be sure to get a holster (so you don't burn your table) and some steel wool, which you should use to clean molten solder off your iron before you put it in the holster. Don't get a soldering *gun* for your synth work, though—it's made for industrial plumbing and ductwork and won't be much help for soldering tiny wires together.

A roll of rosin-core solder

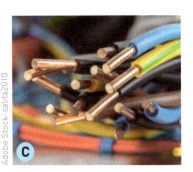

There are many kinds of solder, each with a different delicious recipe. The kind we're looking for is rosin-core, which has a thin tube of pine tree sap running down the middle (Figure B). This sap melts and pulls the metals along with it, helping your solder flow the instant you touch it. Whatever you do, be sure not to buy acid-core solder—that's for plumbers, not electricians.

When shopping for solder, you'll notice that some kinds contain lead (Pb), while others are advertised as lead-free. Lead is a dangerous substance and shouldn't be taken lightly, but in this case, *we recommend using it*. Lead-free solders melt at much higher temperatures and are much more annoying to learn with—trust us. Just be sure to take caution with the smoke it generates. Always solder in a place with good ventilation, preferably with an open window or a fan on.

Solid and stranded wire, 22 AWG (22 gauge)

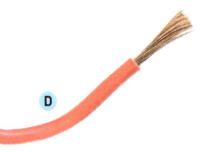

Adobe Stock-salita2010

There are two kinds of wire: solid and stranded. Solid wire looks like Figure C. Stranded wire looks like Figure D.

Solid wire has one solid channel of metal and is surrounded by an insulator. Stranded wire has many channels of metal and is surrounded by an insulator. "22 AWG" refers to the diameter of the wire—this is the size we've found works best for these projects.

While both kinds of wire can be used interchangeably, *solid wire* is designed for breadboarding (where its pointiness and durability make it reliable), while *stranded wire* is designed for situations in which the wire will be moving and bending. (The hairlike fibers of the wire make it amenable to contorting and stretching in a way that would break a solid wire.) We recommend having both kinds around—and while you're at it, get a variety of colors, too. (Color-coding solder makes it easier for you to keep track of stuff.)

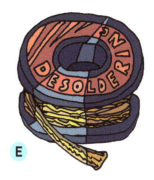

E

A desoldering braid/wick or pump

Both of these are tools that help pick up extra solder if you accidentally put on too much of it or if you need to remove a component later. The desoldering braid (Figure **E**) looks like a little piece of tape with holes punched into it—if sandwiched between the tip of the iron and the solder you want to remove, the braid will absorb the molten metal like a steampunk sponge. A desoldering pump is somehow an even goofier apparatus that creates a small vacuum and sucks molten solder up faster than you can say Jack Robinson.

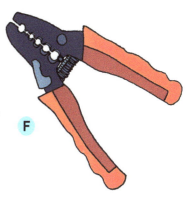

F

Strippers

No, not the pole-dancing kind. Wire strippers (Figure **F**) are useful for removing the insulated layer from the outside of your wires so you can put them in your breadboard or circuit board.

A pair of flush cutters

If you've ever looked at a pair of scissors closely, you'll notice that the two blades kinda sideswipe each other on a diagonal. This results in slanted cuts which are OK for paper but are pretty bad for wires. The word "flush" means "straight up and down"—flush cutters' blades meet in the middle to form a single plane, so they neatly make straight and level cuts neatly at a perfect 90-degree angle (Figure **G**).

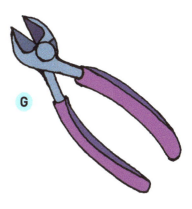

G

A thing to solder to

If we want to transfer our CD4093 oscillator/LM386 amplifier combo to a permanent home, we'll need a board to put it on. For first-time solderers, I recommend purchasing a pack of perma-proto (permanent prototyping) boards. The name may seem oxymoronic, but they're permanent in that they're true circuit boards, but the "proto" comes from the fact that they're laid out exactly like a prototyping breadboard. Take a look:

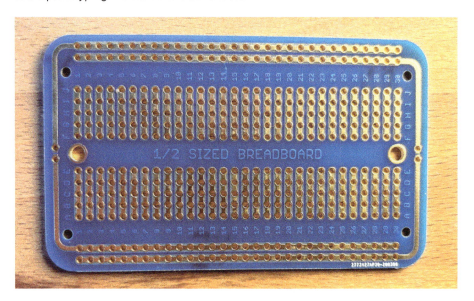

We recommend these boards because they're super logical—if you've made a breadboard model that works, all you'll have to do is rebuild the circuit the same way on the perma-proto.

A pie

For snacking later.

Other items

I promise you won't regret having these items around, despite the initial startup cost. Building electronics is an astonishingly empowering hobby, and not much else compares to playing a finished, soldered instrument you made yourself.

- **Standoffs** are tiny legs you can add to a circuit board to make it look a little tablelike. These are handy for big projects that involve multiple boards, as they let you stack boards like floors in a building. We won't need them for this project, but you might find them handy later on in your synth-building journey.

- *Heat-shrink tubing* resembles a little rubber tube. If you put the tube around a junction with exposed metal, you can heat it up with a lighter or a hair dryer and watch the tube shrink to provide an airtight lock. Heat shrink is great for covering up naked wires and ensuring that they don't accidentally touch later on.

- *A helping hand* is a fancy desktop thing with little gator clips that you can use to grip stuff while you solder. Because not everyone has these, our instructions show you how to get by without one.

SOLDERING SAFETY

Soldering irons are hot and produce fumes that are less than healthy. Most of our safety tips boil down to "Take it seriously."

- **Never, ever, ever, ever, ever, ever, ever, ever, ever, ever grab the soldering iron** *anywhere where metal is exposed*. Instead, grab the handle with the nice, cushy foam bit. The internet is filled with amusing stock photos of people holding soldering irons by the wrong part, which we will include here:

If you hold a soldering iron like these models, you'll be running to the emergency room!

Adobe Stock-auremar & Africa Studio

- Never put a hot soldering iron on the table (unless you no longer want to have a table). Use a holster instead.

- If you drop a soldering iron, don't try to catch it. Let it fall, let the tip break, and buy a new tip. It's not worth the risk of catching the wrong end.

- The fumes that come from solder aren't going to kill you faster than anything else (unless you're soldering forty hours a week), but we can most certainly say they're not *good for you*. Always solder in a place with ventilation. Work with a window open, or next to a fan positioned so it draws the fumes away from you. If you have asthma or another respiratory condition, we encourage you to take this tip especially seriously.

- Don't leave the soldering iron plugged in when you leave the room.

All this being said, in the several years I've been teaching synthesizer workshops, I have seen zero soldering-iron-burn incidents and approximately 10 zillion hot-glue-gun-burn incidents. Based on years of empirical data, I would like to propose Pearson's law of ouch: *The gnarlier the injuries from the tool, the more reverence you'll give the tool.*

This is why so many people are injured by vending machines every year, and yet there are very few nuclear-waste incidents. Or something like that.

PROJECT: COMPLETING THE PHOTO THEREMIN

INSTRUCTIONS:
Part 1: Attaching Wires to Speakers
Let's start our soldering journey by attaching two long wires to the back of a speaker.

1. Cut two long sections of stranded wire, and strip the insulation off a half-inch or so of each tip (Figure **A**).

2. Thread these wires through the two leads on the back of a speaker. Twist them around so they won't fly out when you try to solder them (Figure **B**).

3. With your soldering iron on and up to temperature, touch the end of your roll of solder onto the tip. In a second, you'll see the solder melt and glob up on the tip of the iron. This glob of solder, interestingly enough, won't be the solder that goes onto the wire-speaker junction. It's just there in order to conduct heat more efficiently from the iron to the solder. *If you do not put this glob on the tip, you'll find your solder doesn't melt super quickly.*

   We call the process of covering the tip of a soldering iron with molten solder *tinning* (Figure **C**). (Note the noxious white smoke coming from the molten solder.)

MATERIALS:
- A few feet of stranded wire
- Wire strippers
- A speaker
- Solder, a soldering iron, a desoldering braid or pump, steel wool, a holster
- All the materials needed for our Hello World oscillator (page 59)
- All the materials needed for our Power Amplifier (page 75)
- A blank perma-proto board (or circuit board)

A

B

C

4. Position the soldering iron on top of the metal junction between the stranded wire and the speaker lead. (Solder flows best onto hot things.) Then, using your other hand, touch the end of the roll of solder onto that junction. If all goes well, the liquid solder will start flowing onto the junction (Figure **D**).

5. Once enough solder has accumulated for you to see a little metallic bead of molten metal (Figure **E**), pull your soldering iron away from the speaker. Dip the tip of your iron in steel wool to remove residual solder (it'll prevent the tip from getting gross and unusable in the future). Finally, put the iron back into the holster, where it won't fall over.

6. Now we need to test the speaker. Wait a few seconds for the solder to cool and solidify. When it's sufficiently cold, try lifting your speaker with each wire—the connections should be strong enough to physically lift the whole speaker off the table. If your wires break off, cut off the tips and try soldering again.

7. When both wires are attached and the connections are solid, gently and quickly tap the two wires to the two leads on a 9 V battery. You should hear the speaker "pop" in or out, just like how we tested our speakers back in chapter 2.

8. If the electrical test fails, it likely means one of the following:
   a. You have a cold solder. This means you have a joint that's structurally sound but electrically disconnected. You can often spot these by their milky (and not metallic) appearance. Just use your desoldering braid or pump to pull up the extra solder, and try conjoining the two again.
   b. You have a broken speaker or a dead battery.
   c. You have forgotten what a pop sounds like.

Part 2: Soldering the Circuit Board

For our first build in this project, we'll be modifying our oscillator-and-amplifier setup. We'll be making a photoresistor-controlled oscillator, much like we did at the end of Chapter 3, but we'll also add a few more helpful features:

- **An amplifier, so you can hear your oscillator without having to plug it in**
- **An on/off switch, so you don't have to manually detach the battery when music time is over**
- **An indicator LED, so you can quickly tell if the unit is getting power**

1. Build a CD4093 oscillator and amplifier on a breadboard, as described in the last two chapters (see pages 69 and 75), but include a photoresistor in place of a potentiometer (Figure **F**).

2. Let's add an indicator LED that will show us when the circuit is receiving power. This will make it clear if our battery has died. Add a 330 ohm resistor and an LED to a spare row on the breadboard, making sure that the shorter leg goes to the battery minus row, and the longer leg attaches to the resistor. Make sure your LED shines when the battery is connected and turns off when the battery is disconnected (Figure **G**).

F

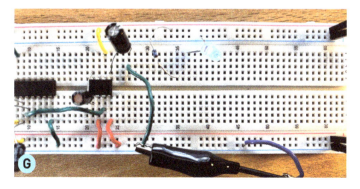

G

3. Our circuit will ruin your home life unless we include an on/off switch. For this project, I'm using a momentary-on tactile, or tact, switch with a big blue push button. You'll learn more detailed information about these in our chapter on schematics, but basically, pressing the button connects the two pins diagonal from each other (Figures **H** and **I**).

    Our goal is to have this button connect and disconnect the battery. One way to do this is to attach the battery minus wire to one of the pins on the button. Then, attach the *diagonal pin* to the battery minus row. What does this do? When I press the blue button, our battery's ground connection connects to the circuit and our oscillator squeals to life (Figure **J**).

4. Once we have our on/off switch and indicator light, we're ready to start the process in earnest. Take a deep breath, and *gather new components to make a totally new oscillator*.

    It's very tempting to just pull the parts out of your breadboard and solder

them in one by one, but we wholeheartedly discourage this! Whenever you want to solder a circuit together, make sure you have a working prototype already built. That way, if there's an issue with your build, you can easily figure out where the problem is by comparing the circuit with your prototype (Figure **K**).

5. When building a breadboard prototype, it's easy to put your components in place in any order. On a circuit board, it's best to put the pieces in by order of *flatness*. Trust us—this will make it easier. Let's start out with the flattest of them all: our integrated circuits.

When putting ICs into a circuit board, some people like using *IC sockets*. These are little seats, built for any size of IC, that are soldered on a circuit board. The actual ICs are then popped into these little thrones and can be easily plucked out and replaced in case they ever break. IC sockets are optional—you can always solder the IC directly onto the circuit board—but this makes them harder to swap out if needed.

However, for the sake of the images in this chapter, we're gonna ignore IC sockets. (But we'll still include IC sockets in the written directions.)

Let's place one IC socket on the left-hand side (for our CD4093) and one on the right-hand side (for our LM386). It'll look more or less exactly like a breadboard layout (Figure **L**).

Tape your ICs so they remain on the board (Figure **M**). This will make it much easier to flip your perma-

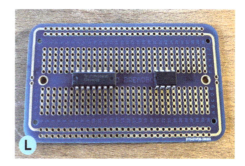

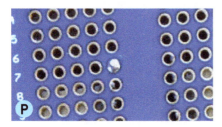

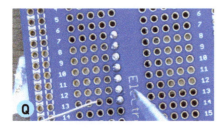

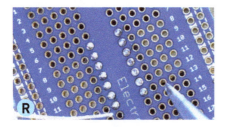

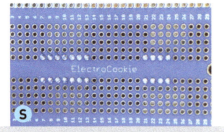

proto board over—do that now! You'll be able to see the tiny little feet of the ICs just barely poking through the bottom (Figure **N**).

We're now going to slowly and carefully solder each of these pins to its respective conductive trace. To do this, we'll put a very, very small amount of solder on each of the pins one at a time (Figure **O**).

Put a little glob of solder on your tip, position your tip on one of the sides of an IC leg, and touch the end of your solder roll to the other side. Within a few seconds, the solder should melt and flow onto the leg and trace. Immediately remove the iron and solder once enough has accumulated. Repeat the process for every leg. This might start off slow, but once you get comfortable, you'll be able to manage a quick, second-long tap on each leg (Figure **P**).

Start with one pin, and make sure you see a shiny little blob of solder holding that pin in place (Figure **Q**).

Then continue down one side of the IC, securing each of the pins into place.

Now do the other side of the IC like so (Figure **R**).

Be careful about the amount of solder you use. If you put too much on one spot, it might accidentally bridge to somewhere else—for example, it's not hard to accidentally join two adjacent legs or connect a leg to the incorrect trace on the board. You can always use your desoldering braid or wick to remove the excess, but you'll also notice that a fine soldering tip is the perfect length to draw lines between traces that will remove all excess solder (Figure **S**). What a handy little trick!

6. Now that our ICs are in place, let's attach some wires to make our internal connections. Cut several small bits of solid wire for this, as we'll need a few.

7. Connect the wires as follows:

On the 4093:
- Pin 1 and 14 to battery plus
- Pin 7 to ground
- Pin 3 to pin 3 of the LM386

On the 386 (Figure **T**):
- Pin 2 and 4 to ground
- Pin 6 to battery plus

All these wires are gonna fly out when we flip the board over, so let's stick them in place with some tape (Figure **U**).

And now it's time to solder the connections. Tap the soldering iron to one side of the little wire, and tap the solder to the other side and wait for it to melt. These connections can be done quite quickly—just a few seconds for each one (Figure **V**).

Time for the resistor for the LED. Attach a resistor to the perma-proto board and flip the board over (Figures **W** and **X**).

You'll probably notice that the resistor's legs stick way out. Once you solder them into place, use your flush cutter to get rid of all that extra wire (Figure **Y**).

8. On to the capacitors! Put all three into place, and bend their legs a little so they don't wiggle around too much.

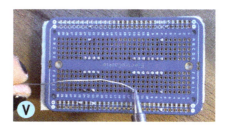

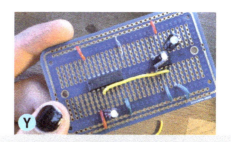

9. Finally, make sure there are no breaks in the battery connections on your perma-proto board. Some board manufacturers put breaks in so you can have different voltages on different rails. We want our top and bottom battery rows to be connected to each other, so I'll add a little bit of red wire to bridge both of these connections (Figure **Z**).

You'll notice that we did not attach our LED, switch, or photoresistor—that's because these three parts will be on the exterior of the enclosure (Part 3). We'll be drilling holes to put them in place and then gluing them to the enclosure before we solder them. For now, I'll call these peripherals, because they sit on the outside of the enclosure.

In order to put our LED, switch, and photoresistor far away from the circuit board, we'll have to add some extension wires. Here's my recommendation for how to do this:

a. Grab a long wire and strip the ends. Put one end underneath a book—or, if your studio looks anything like mine, a stack of Dogbotic stickers.

b. With one end of the wire sticking up toward you, melt a bit of solder onto the tip by touching the iron and the solder to either side of the bare wire. This will result in a little bit of solder sticking to the wire, which will quickly harden. Do this to another wire as well—we'll need two wires to extend the legs of our photoresistor (Figure **Aa**).

c. Line up the leg of the photoresistor parallel with one of your newly gunked-up wires. The goal here is to melt that blob of gunk as the wire touches the photoresistor so molten solder flows and binds the two wires together (Figure **Bb**).

d. Once you have extension wires on the photoresistor, do the same for your LED and your tact switch. Make sure the connections are strong—these are going to be the easiest connections to break.

Part 3: Designing an Enclosure

In order to make our photo theremin an actual instrument and not just a science fair project, we'll need to put it in a box.

Let's start by finding a good enclosure. Remember, fancy materials do not always make good enclosures! Metal, while classy, is pretty hard to cut holes into. Cardboard, while cheap, can be transformed into virtually any shape. We found this cool-looking translucent case to use for this demo (which would have been really cool in 1996) (Figure Cc).

1. The first step in our enclosure's journey is to consider which parts need to be sticking out into the outside world. In this case, all we really need is our photoresistor and our button. Because this case is translucent, we can leave our indicator LED on the inside. (We thought it looks cooler that way.)

2. Drill some holes into the part of the enclosure where you want those pieces to stick out (Figure Dd).

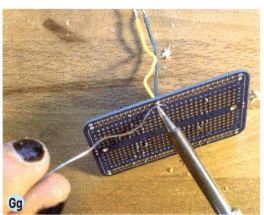

3. Next, we'll drill some holes in the front for the speaker (Figure Ee).

4. While respecting the glue gun and Pearson's law of ouch, carefully extrude a ring of glue around the rim of your speaker and affix it to the underside of the surface (Figure Ff).

5. The next step is to solder on the indicator LED. Because this is going inside the box, I don't have to worry about threading it through anywhere. I'll solder the shorter leg's extension wire onto the battery minus row and the longer leg's extension wire onto the resistor (Figure Gg).

6. Now, let's take care of the remaining peripherals—the photoresistor and the on/off button. Thread these components into the box from the outside and affix them into place with a dab of hot glue. Don't solder these components first—otherwise, you won't be able to get them to the outside of the enclosure (Figure Hh).

7. Solder the photoresistor into place. The legs should attach to pins 2 and 3 of the CD4093.

8. The final, and perhaps most confusing, step is to connect the button. Solder the black wire on your 9 V battery connector onto one of the button's extension wires. Solder the other button extension wire to somewhere on the battery minus row on your circuit board. Finally, solder the red wire on your 9 V battery connector to somewhere on the battery plus row on your circuit board.

Ready to test it?

One press of the button . . .

. . . and your LED will light, your oscillator will ring, and your friends will say, "Please turn that off."

   Is this the world's greatest musical instrument? No. Is it even a useful musical instrument? I don't know. But the important thing is that *you made it*. You took a whole bunch of metal components never intended to be a musical instrument and conjured something meaningful out of them. Once you've made your first instrument, it'll become much simpler to make your second, and so on and so on. Do not keep sitting here reading—make one! You'll be very happy you did. 🎵

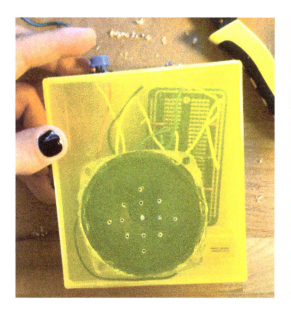 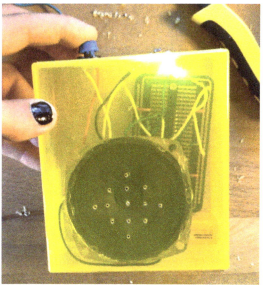

Musical Innovation: Saltina

Artist: Maisy Byerly

About: Maisy Byerly (who illustrated this book) enjoys making friends of many shapes. Saltina is a friendly CD4093-based oscillator soldered into a happy, salty home. In addition to sporting an on/off switch and stylish can-based reverb unit, Saltina's two antennae control pitch and pulse width, allowing her to sing with the birds before they try to eat her.

After completing an oscillator circuit, you might say to yourself, "Self, how does this oscillator work?"

Here at Dogbotic Labs, we subscribe to an ironclad dictum of music technology: You don't have to know how anything works in order to make art with it.

That being said, if you *do* know how the oscillator actually works, you won't only be better equipped to fix it when it breaks, you'll also have a surprisingly nuanced understanding of how your computer works. Let me explain.

The illustrations on the right show a device called a *relay*. It's pretty straightforward: The principal part of a relay is a little coil of wire that turns into an electromagnet when current is sent through it. When the electromagnet turns on, a little switch connects between both sides of the relay and a tiny click is heard. When the electromagnet is off, this little switch disconnects.

The cartoony images to the right show a relay for what it is—an electrically-controlled switch. When we send current to the electromagnet, we complete a circuit. It's deviously simple.

If we hook up the relay mains to a lightbulb and send our oscillator to the electromagnet, we'll make the lightbulb switch on and off, hearing a little click on every state change. This is precisely how the blinkers on a car work! A relatively low-voltage signal is used to electromechanically switch on and off a much-higher voltage source. Most thermostats also work on relays. It is far too dangerous, however, to quickly switch a 120 V load on and off—a relay makes sure a tiny signal can influence a bigger signal.

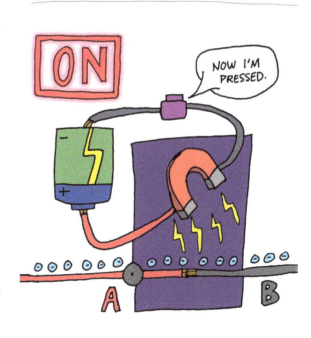

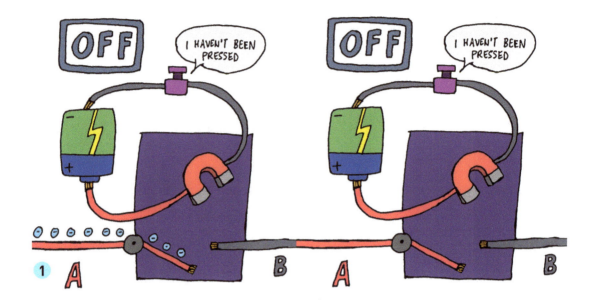

I promise it gets more interesting.

Let's now take two relays and attach them together back-to-back, like you see in Figure **1** .

If we press the button on one relay or another, electrons won't get through. Only when you engage *both* relays can electrons flow all the way through. At first glance, this is pretty uninteresting. But on second glance, you might notice you've built a little piece of electronics that can do something quite existential: it can understand logic.

What?

Well, yeah. You've just built what computer scientists call an *AND gate*: a device that outputs current if—and only if—both of the inputs (electromagnets) are switched "on." If A *and* B are switched on, our output light will glow. Let's press both of the buttons on our relays. Notice how both of the drawbridges slide up into place, letting electrons flow all the way across our wire (Figure **2**).

We can make this relationship a little clearer by writing out what computer scientists call a "truth table", a little chart that shows us which combinations of inputs will produce a particular output. Let's make a chart that covers all our possibilities:

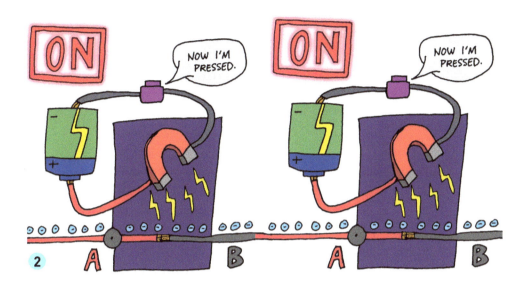

Relay #1	Relay #2	Will Electrons Make It Through?
Button not pushed	Button not pushed	No
Button not pushed	Button pushed	No
Button pushed	Button not pushed	No
Button pushed	Button pushed	Yes

You've probably noticed that this chart could be rewritten with simple yes or no answers. We can make this a little more computer science–y by writing in the number 0 to represent "no" and a 1 to represent "yes."

Relay #1	Relay #2	Will Electrons Make It Through?
0	0	0
0	1	0
1	0	0
1	1	1

AND Gate

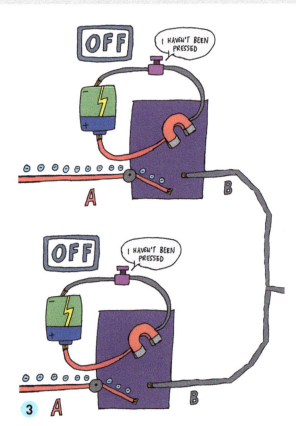

Ta-da! You're now a computer scientist. Easy, right?

Now, what happens if instead of connecting the relays in a long line, we connect them like you see in Figure **3**?

While the AND gate involved two relays being connected *in series* (one after the other), this configuration is what we call *parallel* (each relay offers a separate, parallel path for current). We call this circuit an *OR gate*, because if A or B is switched on, electrons will make it through. In fact, the only way electrons won't flow through is if neither relay is turned on, like in the illustration.

We can make a truth table for the OR gate like so:

OR Gate

Relay #1	Relay #2	Will Electrons Make It Through?
0	0	0
0	1	1
1	0	1
1	1	1

There is one other logical operation we'll talk about in this chapter, and that's an operation called *NOT*. As you might expect, NOT produces the opposite of whatever the input is. If you input an "on" to a NOT circuit, it will respond by turning your lightbulb off. An "off" message to a NOT circuit will turn your light bulb on. Whatever you tell it, it does the opposite.

This may sound rather complex, but you can make a NOT gate by essentially building a relay in reverse. We call a "relay in reverse" an inverter because the output will be the *inverse* of whatever the button's status is. If the button is pressed, no electrons will come out, and vice versa (Figures 4 and 5).

Take a look at the inverters on the right—notice how, unlike a relay, this device is normally closed. Only when the electromagnet is turned on does the switch flip and current *stops* flowing.

A NOT gate is more commonly called a "logic inverter" because it presents whatever the inverse of the input is. This results in the world's least-interesting truth table:

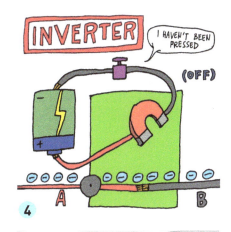

4

Inverter	Will Electrons Make It Through?
0	1
1	0

NOT Gate (inverter)

If you recall from Chapter 3, the CD4093 is a circuit called a NAND gate. NAND is a mishmash of the words "not" and "and." If we replace the button on an inverter with *the input of an AND gate*, we will have a circuit that produces the exact opposite of whatever the AND gate would (Figure 6).

This results in the following truth table, which is the

5

6

opposite of the AND gate. All the zeroes have become ones, and all the ones have become zeroes.

Relay #1	Relay #2	Output
0	0	1
0	1	1
1	0	1
1	1	0

NAND Gate

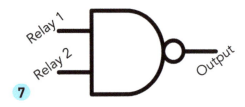

7

A nice thing about computer architecture is that once you understand how the fundamental pieces work, you can abstract the minutiae away into something more digestible. Instead of this giant mess of boxes, let's simplify by giving the NAND gate its own shape (Figure **7**).

You'll notice, despite its apparent complexity, that the NAND gate really has only two inputs and one output. When both of the inputs are high, the output turns low. In any other state, the output is high.

Moving on back to our CD4093 (Figure **8**), you'll notice that there are not one, not two, not three, but four NAND gates! Our first oscillator used only one of them, but in the next chapter, we'll be exploring all sorts of unexpected sounds you can generate by having the oscillators talk to each other.

We can now think of pins 1, 2, and 3 on our 4093 IC as the three pins on a NAND gate. Pins 1 and 2 represent the relays, and pin 3 is our output. In our oscillator, we've attached one NAND gate input permanently to 9 V—or, in truth-table-speak, a 1. We've attached the other input to a capacitor,

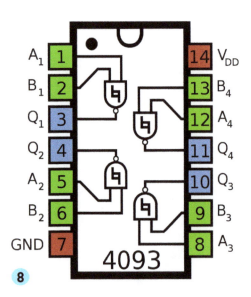

8

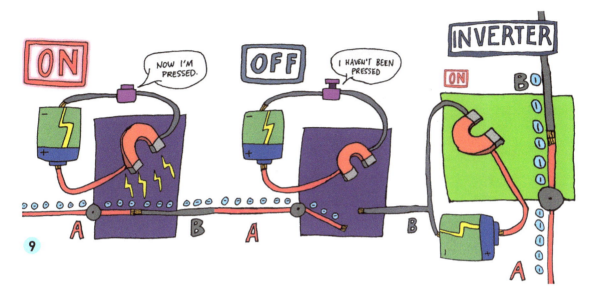

which drains to the battery minus row. We've also attached the output (pin 3) to one of the inputs (pin 2).

Let's slow down time and walk step by step through what our NAND-gate oscillator is really doing.

Step 1:

The instant the battery is connected, our NAND -gate configuration looks like this:

1 0 1

Input 1 (pin 1, a.k.a. relay button 1) is high, and input 2 (pin 3, a.k.a. relay button 2) is low, and thus the output (pin 3, a.k.a. our inverter output) is high. Immediately, current starts pouring out of pin 3 (Figure **9**).

Step 2:

That output from pin 3 starts flowing through the potentiometer, directly into the capacitor. Over the next few milliseconds, the capacitor starts filling up with charges, causing the voltage at pin 2 to rise.

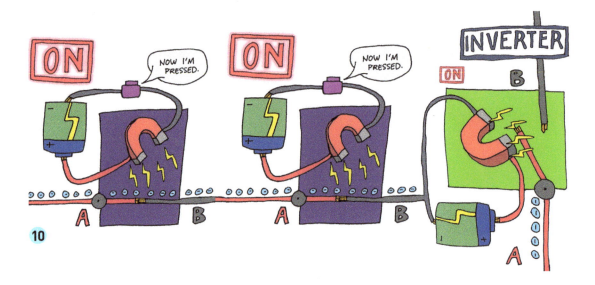

Step 3:

Eventually, the capacitor has enough charge in it for input 2 to "turn on." Now, both of our inputs are high! Suddenly, when pin 2 goes high, the output turns off (Figure 10).

Step 4:

Because of the suddenly dropping voltage at pin 3, all the charges from the capacitor have to flow somewhere. They escape the only way they can— backwards through the potentiometer toward the output!

Step 5:

As the charges run out of the capacitor, pin 2 will eventually "turn off," turning pin 3 on and starting the whole cycle over again.

What's so nifty about this NAND-gate oscillator is that you're sending electrons back and forth like some sort of electrical seesaw. By causing the voltage to drop in one place, and then another place, and then the first place again, you'll force tremendous amounts of electrons to wiggle in a musical pattern.

The NAND gate we've hooked up is outputting a pulsing alternation of 9 V and 0 V. If we were to graph this voltage over time, we would make a pretty boring graph that looked like a series of squares. Because of this shape, we call the sound we get from this oscillator a "square wave." There are many other kinds of waves too—we'll talk about them later.

PROJECT: POLYPHONIC SQUARE WAVES

MATERIALS:

- An LM386 amplifier, built on your breadboard, or a separate amplifier with audio cable
- A 4093 integrated circuit
- Assorted jumper wires (it helps to have colors that come in pairs)
- 2 x potentiometers (100 kΩ is fine)
- 2 x 0.1 µf capacitors
- 2 x resistors (10 kΩ or so)

Polyphony comes from Greek. *Poly* means "many," and *phony* means "a person who pretends to be someone else."

Just kidding.

Polyphony means "many voices." In the musical-instrument world, a "voice" is an audio pathway that can make a single note. When you play a piano, you can press every key at the same time (if you have very big hands) and hear all eighty-eight voices. On a trumpet, however, you'll find it pretty difficult to get more than one note out at the same time. A trumpet, unlike a piano, has only one voice—it's *monophonic*.

Let's turn our mono synth into a true poly synth by adding a second tuneable voice. Let's make use of those additional NAND gates on the chip. This means you're no longer stuck with one measly square wave—you can now have four!

Taking a look inside our 4093, we see four NAND gates, each clustered around a trio of pins. Our first NAND gate (and, thus, our first oscillator) uses pins 1, 2, and 3. Our other oscillators can use pins 4, 5, and 6; pins 8, 9, and 10; or pins 11, 12, and 13. Note that the NAND-gate configuration is mirrored. When you build additional oscillators, you have to make sure you remember which pins are the inputs and the outputs.

INSTRUCTIONS:

1. Build two identical oscillators on the breadboard. Here, we've built one using pins 1, 2, and 3 and another using pins 8, 9, and 10 (Figure **A**).

A

B

2. Now, let's hook those oscillators up to our amplifier to hear those two tones sing together. Logically, to add those two signals together, we just have to add them! However, you'll notice that if we take both outputs of our oscillators (pins 3 and 10) and send them to our amplifier via jumper wires (Figure **B**) . . .

. . . we don't hear anything.

3. However, if you put each oscillator through a small resistor (10 kΩ or so) and then mix them, you'll hear the two tones perfectly. Notice how our outputs (pins 3 and 10) now go to jumper wires, which in turn go into resistors. The two resistors then meet up in a common row, where they are sent to the amplifier (Figure **C**).

Why Does This Happen?

Without the resistors, when current comes out of one oscillator, it actually *flows back into the output of the other oscillator* instead of to the amplifier. Putting resistors in place gives our current two options: flow easily to the amplifier or take a much more arduous path through a *second* resistor to get to the oscillator output. Going through two resistors is twice as much work as going through one resistor, so the current literally takes the path of least resistance. Nifty, huh?

We can expand on this madness by adding photoresistors in place of our potentiometers and building another two oscillators. Now we have a nightmarish drone machine that's sensitive to light!

Fun project idea: Build a photoresistor-controlled poly synth, hang a flashlight from above on a long string, and set the pendulum in motion.

If you play around with different resistor values as you mix your signals, you'll find you can change the relative volumes of the different notes.

- An LM386 amplifier built on your breadboard, or a separate amplifier with audio cable
- A 4093 integrated circuit
- Assorted jumper wires (it helps to have colors that come in pairs)
- Two potentiometers (100 kΩ is fine)
- A ~22 µf capacitor and a ~0.1 µf capacitor

PROJECT: TREMOLO

Polyphony is fun and all, but where we can really start cooking with gas is when we *use one oscillator to control another one*. What do I mean by this? Let's say I have two square waves—one vibrating at audio rate (making an annoying whining sound) and another one vibrating at subaudio rate (making a low-frequency clicking sound). What if I used that slow-moving signal to turn the audio signal on and off? Instead of hearing "*beeeeeeeeeeeep*," we'd hear "*beep-beep-beep*." This technique of playing a sound and stopping it regularly is called *tremolo*. Picture a violin player shaking the bow back and forth across the strings to produce a stuttery rumble, and you'll have a good idea of what tremolo is.

As it turns out, it's fantastically easy to accomplish this with the 4093. With every NAND gate, both of the inputs have to be tied high (hooked up to 9 V) for the output to make a sound. If one of those inputs were to be manually removed and connected to the battery minus row, we'd hear that oscillator temporarily stop sounding. To produce a beeping sound, we'd have to manually pull out that input cable and alternate it between 9 and 0 volts. Fortunately, we don't have to do this manually—we can send another square wave (which alternates between 9 and 0 volts) instead!

INSTRUCTIONS:

1. Begin by building two oscillators on your 4093. One is oscillating at audio rate (pins 1, 2, and 3) and the other at subaudio rate (pins 8, 9, and 10). The only distinction between the two oscillators is the capacitor—the larger capacitor makes the IC oscillate at a slower rate (Figure A).

2. Now, watch as we take the output from the subaudio oscillator and send that signal to the input of the audio oscillator. Notice how the *output* from pin 10 now goes into one of the *inputs* at pin 1 (Figure B).

Our output is now an audio-rate oscillator turning on and off at the frequency described by our second oscillator. Continuing with our violin analogy, the vibration of the string is described by our audio oscillator, and the vibration of the violinist's hand is described by our subaudio oscillator.

You will notice that although we are using two oscillators, we have only a single voice. (Again, our instrument is monophonic.) Multiple oscillators can be used to create a single voice!

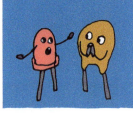

- An LM386 amplifier, built on your breadboard, or a separate amplifier with audio cable
- A 4093 integrated circuit
- Assorted jumper wires (it helps to have colors that come in pairs)
- 3 x potentiometers (100 kΩ to 250 kΩ are best)
- A 22 µf capacitor, a 4.7 µf capacitor, and a 0.1 µf (aka 100nF) capacitor

PROJECT: THE ELECTRO-CRICKET

This goofy little project was one of the first we developed for the DIY Synthesizers workshop. It's simple and charming and holds its own as a piece of art. The Electro-Cricket really does sound like a cricket, and you can turn any backyard into a lovely insect-inspired installation with just a few of these.

The difference between the Electro-Cricket and the tremolo circuit we just built is, what if we had a *third* oscillator that turned the second oscillator on and off? We could derive a pattern that sounds oddly like a cricket: *"beep-beep-beep* [pause] *beep-beep-beep* [pause] *beep-beep-beep . . ."*

We can build a cricket circuit with only some mild modifications to our existing oscillators.

INSTRUCTIONS:

1. Let's begin with a lone CD4093 and an LM386 amplifier (Figure **A**).

2. Attach pins 8, 12, and 14 of your 4093 to +9 V. Attach pin 7 to ground, also known as battery minus (Figure **B**).

3. We'll now need to make three internal IC connections (Figure **C**). These are:
   - Pin 10 connected to pin 1
   - Pin 11 connected to pin 5
   - Pin 13 connected to pin 3

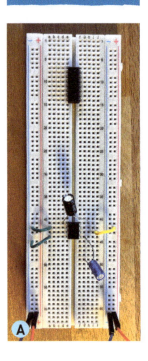

A

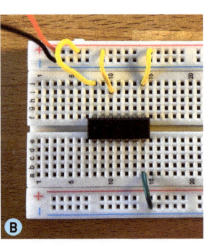

B

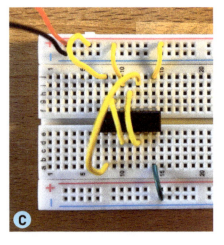

C

4. Now it's time for our capacitors. We'll need three values for this circuit (Figure **D**):
   - A 22 µf capacitor between pin 9 and ground
   - A 4.7 µf capacitor between pin 2 and ground
   - A 0.1 µf capacitor between pin 6 and ground

5. Add three potentiometers for our three oscillators. We recommend using potentiometers between 100 k ohm and 240 k ohm for best effect (Figure **E**). These potentiometers go:
   - Between pins 9 and 10
   - Between pins 2 and 3
   - Between pins 4 and 6

D

E

6. Finally, attach the output from pin 4 to the input of our LM386 amplifier. Attach the 9 V battery and a speaker, making sure that your speaker tabs go to the output capacitor of our amplifier and ground (Figure F).

Most likely, your circuit won't immediately sound like a cricket. Turn the potentiometers, and with a little fiddling, you'll get something that sounds shockingly insectlike. You can think of these three potentiometers as three parameters of a cricket's chirp. The first oscillator, which determines the pause length, represents how much the cricket cares to speak. The second oscillator represents the temperature, which proportionally increases with a cricket's chirp rate. The final oscillator represents how quickly the legs are rubbing together. Pretty nifty, huh?

A question we get a lot with this circuit is, "What is that fourth NAND gate doing? We have only three oscillators, yet we're using all four NAND gates!" The answer: We're using the fourth one *as an inverter* in order to flip our control signal "upside down." Without this inversion step, our cricket won't pause for a moment of silence—instead, it will let out a long, sustained tone before beeping again. By inverting that signal, we can make those sustained tones pauses—it's simply sending out a 0 instead of a 1.

I heartily recommend building several of these and sticking them in someone's backyard. It's a really beautiful project and a fully self-contained art piece as is!

PROJECT: THE UNDERTONER

Every once in a while, along comes a circuit so simple yet profound that it makes you reconsider the very foundations of culture.

More on that in a minute.

The Undertoner's circuitry is really no different from the tremolo or Electro-Cricket projects we just described (all of these projects operate on the principle of one square wave turning on and off another square wave). The gimmick here is that this one knows music theory—it can play you a scale.

Somehow, the brainless handful of metal wires you're about to construct "knows" how to play music and never plays a wrong note. Not only will we explain how this marvelous musico-electronic quirk works, but we'll also walk you through a variation in which flickering light composes melodies before your very ears! Just in time for your next candlelit soirée.

INSTRUCTIONS:

1. Put your 4093 IC on your breadboard and connect your power rails. Attach the power pins: 7 to ground (0V), 14 to +9V (Figure **A**).

    Next, connect pin 12 and pin 8 to +9V (Figure **B**).

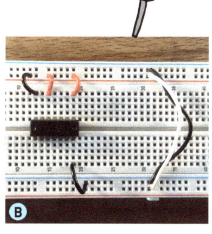

MATERIALS:

- Assorted jumper wires (it helps to have colors that come in pairs)
- 3 x potentiometers (100 kΩ is good)
- 2 x light-dependent resistors (LDRs)
- A few LEDs
- A few fixed-value resistors (anywhere between 330 Ω and 1 kΩ)
- A 4093 integrated circuit
- 4 x electrolytic capacitors (we recommend values of 0.1 μF, 1 μF, 10 μF, and 100 μF or above)
- A breadboard
- A 9 V with battery terminal connectors
- Gator clips
- An amplifier (guitar amps or old computer speakers are great!)
- A cable to plug into the amp

C

2. Make the internal chip connections (Figure C):
   - Pin 1 connects to pin 10
   - Pin 2 connects to pin 4
   - Pin 5 connects to pin 6

3. Put the long, positive leg (the anode) of a 1 μF capacitor into the row shared with pin 3. Put the short, negative leg (the cathode) into the row shared with pin 6 (Figure D).

   Our next capacitor is of the 0.1 μF flavor. We'll want to put the positive leg into the row shared with pin 9 and the negative leg to ground (Figure E).

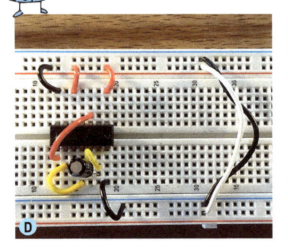

D

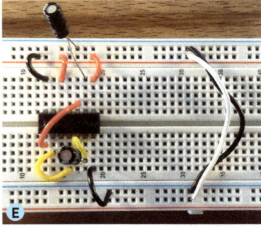

E

Our final capacitor is a 10 µF, which sits between pin 13 and ground (Figure **F**).

4. We'll attach our first potentiometer between IC pins 11 and 13. One of the wires will connect to the nose of the potentiometer. The other wire can connect to either of the cheeks. Which of the cheeks you connect will simply determine the effect of turning the pot one way or the other (Figure **G**).

    The second potentiometer is connected between pins 9 and 10 (Figure **H**).

    The final potentiometer is the trickiest. You'll need to connect one cheek

of the potentiometer to pin 6 and the nose to pin 11 (Figure **I**).

And take a deep breath because we're done. Now for the moment of truth!

5. We can listen to this circuit via either an LM386 amp or an external amplifier (Figure **J**).

Twist potentiometer number 2—depending on which way you turn it, you should hear a gradually upward- or downward-gliding pitch that periodically resets (i.e., jumps back down or up) before starting its ascent or descent again.

Adjust that pot so that you get something you would call a relatively high pitch. Now, head over to potentiometer 3 and gradually start turning it. You should hear—drumroll—a collection of pitches, one after another, that relate to one another in a mysteriously harmonious way.

How mysterious, indeed! If you're a budding synthesist, you might recognize this scale from another classic beginner project: the perplexingly named Atari Punk Console. What could be behind such a device? Well I'll tell you . . .

. . . right after these hot tips!

VARIATION 1: THE ATARI PUNK CANDLE

If you're the experimental type, now would be a good time to swap out the pot on the right for a photoresistor. This way, you can control the transitions between notes with changing light intensity! Grab a candle, enter a dark room, and see if you can find the sweet spot where the candle's flickering makes your Undertoner arpeggiate like a late-'70s disco hit.

VARIATION 2: THE GATING OSCILLATOR

Our fourth (unused) NAND gate seems a little sad out there on its own. Let's turn it into an oscillator and use that to add a "stutter" effect to our already-impressive circuit!

Build an additional square wave oscillator from that unused NAND gate, and use that square wave output in place of the ground connection at pin 2 of RV2.

VARIATION 3: THE VACTROL ARPEGGIATOR

Instead of using a NAND-gate oscillator to stutter our synth, what if we could get it to slowly rise and fall, arpeggiating our scale? This variation is quite similar to the Atari Punk Candle, but with a twist! Instead of using a candle to "compose" your melody, use your spare NAND gate to make an oscillator that flashes an LED on and off. Position the LED right next to the photoresistor so the maximal amount of light is hitting the photoresistor. Now, as your light moves from "off" to "on," you'll hear the circuit quickly cycle through the notes in the undertone series.

Why Does This Happen?

The Undertoner is, in a word, baffling. A few logic gates thrown together somehow give us a collection of pitches that, no matter how hard they try, simply cannot play "out of tune." How in the world do a bunch of subatomic particles understand a musical phenomenon with such deep cultural roots? Let's break down this circuit a little bit to make sense of it.

First, let's talk about how to make a NAND gate oscillate. In talking about NAND gates, it helps to make a truth table that shows us what the output of our NAND gate will be when our inputs are on (represented by a 1) or off (by a 0).

NAND Input 1	NAND Input 2	Output
0	0	1
0	1	1
1	0	1
1	1	0

As you can see from the truth table, if we "tie high" one of the inputs to our NAND gate (i.e., connect it to the positive battery terminal), we effectively build an inverter (a logic 0 gets a 1, and a 1 gets a 0). If we connect the other input of the NAND gate to the output, we create a *feedback loop*.

By connecting a capacitor to the so-called feedback input, we add a receptacle for electrons that delays the amount of time it takes for a change at the output to be registered at the input. This is because the input waits for the voltage at its pin to reach a certain threshold before defining its state as 1 or 0, and the capacitor, as it slowly fills, sets the timeline for this process. A bigger capacitor, or a bigger bucket, means a longer time between cycles, i.e. a slower frequency and a *lower* pitch.

By connecting a potentiometer between the output and input, we can control this frequency, slowing down or speeding up the amount of time it takes for the capacitor to fill up. In other words, the potentiometer gives us variable pitch control—and you said logic was boring!

Here is a drawing of the square wave output, a pattern one can describe as "jack-o'-lantern-core."

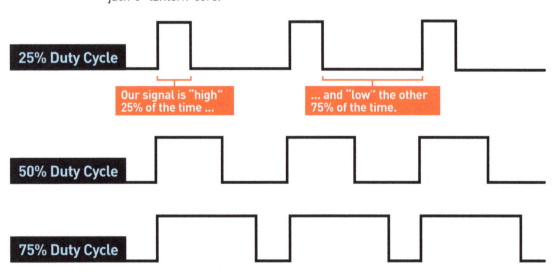

25% Duty Cycle

Our signal is "high" 25% of the time ...

... and "low" the other 75% of the time.

50% Duty Cycle

75% Duty Cycle

Look at the square wave that says "50% duty cycle." This NAND gate's output consists of equal "on" and "off" moments alternating in cyclical repetition. (In other words, the jack-o'-lantern has regularly spaced teeth.)

There's nothing written in the laws of synthesis that says, however, that our on and off moments have to be equal. In fact, we can vary the *pulse width* (i.e., how long the "on" is) with this other potentiometer in your circuit. Take a look at what happens when we vary the pulse width of our square wave to 75 percent or 25 percent of the duty cycle.

Now, take a second to consider what happens if we make the pulse width *longer* than the actual frequency cycle.

In this scenario, the pulse-width control acts also as a signal blocker. As we increase the pulse width past the period of the oscillator frequency, we start blocking every *second* oscillation. It turns out that blocking every other "on" moment halves the original frequency. Halving the frequency, in music circles, is known as "playing the same note an octave lower."

As we continue to lengthen our pulse width, every time it passes a new integer threshold, we'll hear a new fractional subdivision of our original frequency. In this way, we'll progress from frequency X to X/2, X/3, X/4, X/5, and so on.

We call this method of deriving pitches (dividing a number by a series of whole numbers) the *undertone series*. Whereas in nature, a fundamental tone generates an *overtone series*, containing itself as well as all faster (i.e., higher) frequencies produced by integer multiples of its vibrational state, in electronics, the opposite happens.

Luckily for our consonance-loving ears, however, the same intervallic pattern is produced! So, instead of octave up, fifth up, fourth up, major third up, and so on, (i.e., what you get from your ancient Greek monochord), on our breadboard, we get octave down, fifth down from there, fourth down from there, major third down from there, and so on.

Thus, this circuit is deriving the undertone series by dividing our input frequencies by whole numbers! It's a lovely little reminder of just how easily the principles of music apply to the electrical world as much as the mechanical.

You will notice, amusingly, that it's hard to play along with this circuit on most instruments you might have laying around the house. Since the 1700s, the western European compositional tradition (of which instruments like the piano and guitar are emblems) has largely eschewed this mathematically derived tuning system in favor of something called *equal temperament*. This made it easier to have different instruments play together in tune.

While the Undertoner is a "synthetic" instrument, unlike as with a piano or a guitar, the scales it plays are mathematically derived all the same. And when you remember that unlike the undertoner, those acoustic instruments have been unnaturally adapted to suit a different, artificial scale, it almost makes you wonder what the real "synthesizer" is.

THE BIG TAKEAWAY

Way back in Chapter 1, we built a speaker and learned that a voltage signal can be used to represent a sound: Just hook up a voltage signal to a coil near a magnet, and you can make that magnet vibrate in any pattern you so choose. Voltage can be audio.

Now, we've learned that a voltage signal can also be used *as a control signal*. Instead of using voltage to directly move a speaker, we can use voltage to *modulate* a parameter of our audio. For example, we used an oscillator as voltage control in order to periodically turn our sound source on and off. The oscillator that provides the turn-on-and-off information is not actually producing any audio—it's producing a performance, not a sound.

Most commercial synthesizers you find have a feature labeled LFO, or low-frequency oscillator. An LFO is just like an oscillator—capable of different shapes and rates—except that it moves so slowly that you can't hear it. Why give the customer an oscillator they can't even hear? Because it's used as a control voltage. An LFO provides a slow, periodic modulation source.

Moral: audio signals and control signals are made out of the same stuff. Let's move on. ♪

Musical Innovation: Cricket Hotel

Artist: Mei-ling Lee and Jayshing Goolsby

About: Mei-ling Lee is a composer, sound designer, and insect enthusiast who took the Electro-Cricket circuit to the next level: real estate. Made in collaboration with her daughter Jayshing, the Lee Family Cricket Hotel has delighted and confused visitors since 2020. Each of three oscillators is hidden under a little plywood roof, complete with individual indicator LEDs so one can easily see which oscillator does what. Unique to this design is a piezo that serves as the output. Normally, piezos used as speakers result in a dry, tinny sound. Fortunately, this happens to work incredibly well if the goal is mimicking crickets. Try it out yourself!

Musical Innovation: Maneki-Tekno

Artist: Bob Motown

About: Illustrator Bob Motown has made an impressive career drawing cats, pizza, and cats eating pizza. For his DIY synthesizer "final project," Bob created Maneki-Tekno, the world's first meowdular synthesizer—controllable via its arpeggiation arm. The circuitry inside is Bob's personal variant on the Undertoner, so the arm steps blissfully between steps of a just-intoned scale. Bob cleverly hid a potentiometer inside the arm, and the knob serves as the point of rotation. A speaker and battery compartment is hidden in the bottom to make this feline fully portable.

07

Schematics & MASS TRANSIT!

You're now at the point in your electronics journey where you should learn how to read a schematic. A schematic is the crazy-looking diagram you see electrical engineers with—it shows you how to put together a circuit.

So far, we've been showing you how to build everything by using photographs of breadboards. This is totally fine when your circuit looks like the breadboard at top right.

But it becomes a little problematic when our circuits look like the breadboard at bottom right.

This is the point in the book at which our circuits won't get more challenging per se, but they will certainly involve more parts. This will result in your breadboard looking like a giant plate of spaghetti. As circuits become more complex, breadboard photographs become less and less helpful. Tracing wire colors on a photograph is a very frustrating way to make a circuit. Fortunately, schematics were developed for precisely this purpose—they are easy to read and neatly organized and show you exactly what connects to what.

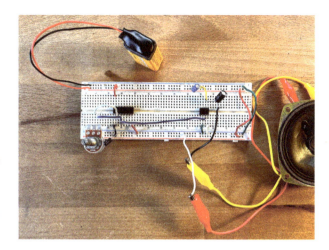

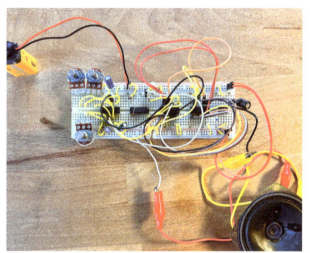

Once you can read a schematic, you can build pretty much anything. Let me say that again for emphasis: ***Once you can read a schematic, you can build pretty much anything***. A lot of people who make electronics don't actually create their circuits from scratch—they find a premade schematic, build a prototype on a breadboard, and, once it works, solder it together. If you want to build, say, a second-order low-pass ladder filter, you can do a good old web search for "second-order low-pass ladder filter + schematic" and one will pop right up.

That's it. Once you can do this, electronics can be paint-by-numbers if that's all you want out of it.

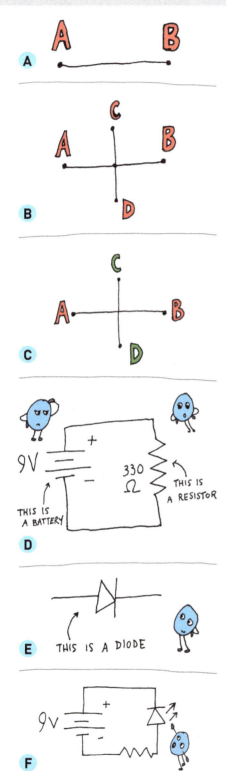

THE HIEROGLYPHS

On a schematic, lines are used to indicate connections. You can think about them as wires.

Figure **A** shows a wire with two points—A is connected to B.

The next drawing shows two wires connected in the middle, where they intersect—you can tell because of that little dot. You'll notice A is connected to B, C, and D (Figure **B**).

If we want to separate these two wires so they *do not touch*, we simply don't add the little dot in the middle. Now, A touches B, and C touches D, but A doesn't touch C (Figure **C**). Make sense? Kinda? Good.

Let's add two components, a battery and a resistor, to our wire. Schematics are intended to be quick to draw and easy to read, so you'll notice a lot of these symbols look nothing like what they represent. Notice how the resistor zigzags, which looks nothing like a real resistor. However, in a poetic sense, the zigzag makes that section a little harder to traverse—imagine having to run through a zigzagging corridor at full blast. The schematic illustration shows you what the thing acts like, not what it looks like (Figure **D**).

Notice we've written out "9 V" next to the battery and "330 ohm" next to the resistor. Every practical schematic will tell you the exact type of component you need, often printed directly next to the symbol.

Let's connect an LED to our circuit after the resistor. An LED is a diode, and a diode is a component that allows current to flow through it in only one direction. As such, the schematic symbol for a diode looks like a one-way street sign, complete with a big, pointy arrow (Figure **E**).

The symbol for an LED is a diode complete with some fun little arrows coming out of it (representing light). Here it is in a circuit with our battery and resistor (Figure **F**).

Notice how the current can flow the direction of the arrow, but in the opposite direction, the current hits a brick wall.

Here are some other symbols worth knowing about:

Ground

Ground, known to its friends as "the negative battery terminal," looks like a cartoonish representation of the ground underneath someone's feet. If you're in a country that is (or at one point was) forcibly taken over by the British Empire, you'll likely see ground depicted as a little triangle (Figure **G**). For some reason, people in North America draw it differently, and we don't know why.

Photoresistor

A photoresistor is just a resistor that can do a cool trick with light. To represent this, we draw a resistor with some little action lines that represent light (Figure **H**).

Potentiometer

A potentiometer, or pot, has three legs—two cheeks and one nose. The middle leg is the nose, which attaches to the wiper, and the outer legs are the cheeks, which attach to the graphite runway the wiper slides along.

   We depict a potentiometer as a resistor and an arrow. The resistor's ends represent the cheeks of the potentiometer. The arrow represents the wiper (Figure **I**).

Capacitor (Electrolytic and Ceramic)

Making a capacitor is like making an insulator sandwich. Between two tasty, tasty slices of conductor is the filling—a delicious insulating material called the *dielectric*. These metal slices are best pals, snuggled up close in parallel, but the dielectric plays referee to ensure that they don't get too cozy.

   Take a look at the schematic symbol for a capacitor, and notice how it looks *exactly* like it sounds (Figure **J**).

   When electric current flows into the capacitor, things get interesting. Charges try to pass from one conductor to the other, but the dielectric prevents that. Electrons, being sassy negatively charged party crashers, will pile up on one of the metal slices, turning it into the ultimate negative hot spot.

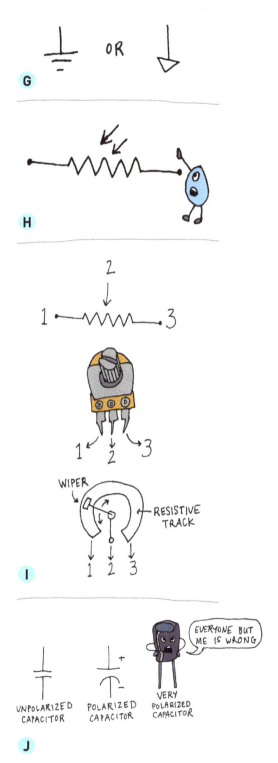

Meanwhile, the other slice, feeling a bit left out, goes positively bonkers as *positive* charges gather on its side. One slice's negative charges pull the positive charges on the other, effectively cramming all those charges closer together. By reducing the amount of space the electrons need, the capacitor has *increased* its standing room capacity. Now, we can fit extra energy in a space that otherwise wouldn't fit it! This is why capacitors act like buckets—they shmoosh charges so closely together that they can fit extra electrons (i.e., extra energy) within their walls.

Some kinds of capacitors, like electrolytic capacitors, are polarized. This means that electric current can flow through them in only one direction. If you accidentally reverse the direction of current or hook up your capacitor backward, it either will not work or—worst case—will **violently explode**. To prevent this, take note if the capacitor schematic symbol has a curvy side; that represents the negative side, while the linear side represents the positive side. Some kinds of capacitors, like ceramic capacitors, don't have polarization—they can be inserted into the circuit in either direction. Those are typically depicted with two linear sides and no curvy bits.

Speakers

The speaker symbol was clearly designed by an artist into realism.

One terminal of the speaker is connected to your audio source. The other terminal is connected to ground.

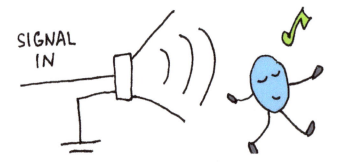

Switches (and Buttons)

A switch is a component that opens or closes a circuit. Most of the time, when you think of a switch, you picture a lever you can twiddle back and forth, like what Dr. Frankenstein would use to kick off his experiments. However, switches can look like all sorts of things—buttons, twirly knobs, or even touchscreens. Switch just means it's capable of rerouting electron flow—which can be done with all sorts of interfaces.

A switch at any point is either open or closed. When a switch is off, it means the circuit is open and no energy is flowing through that part of it. When a switch is on, it means that part of the circuit is closed and electrons are chugging along.

There are two major categories of switches: *maintained* and *momentary*. A maintained switch is what we normally think of as an on/off switch. It keeps the state it is set to until someone flips it back, like when Dr. Frankenstein flips the big, clunky lever on the wall. A momentary switch, however, springs back into its initial position as soon as you finish actuating it—like your computer keyboard buttons. Momentary buttons are ideal for any time something has to be triggered repeatedly.

Push Buttons/Tact Switches

Push-button switches come in all shapes and sizes, from the big and bold to the small and subtle. When you give them a tap, they unleash that satisfying *"click"* that makes you feel like you're in control. Some have fancy LED lights shining through their buttons as an indicator for when you switch it on. (Just connect the LED terminals on the back to power, and you'll be good to go.)

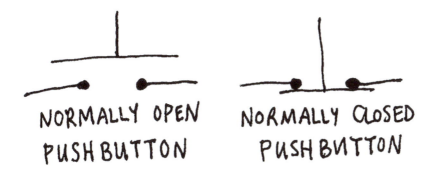

NORMALLY OPEN PUSH BUTTON   NORMALLY CLOSED PUSH BUTTON

Button Matrices

Picture a pocket calculator's array of buttons. When a button is pressed, it pushes down on a wire associated with its row, which in turn lowers onto another wire associated with its column. Your calculator knows which button is pressed because of the specific combination of row and column. If you take apart an old button matrix, you'll find you can wire anything into the rows or columns independently. To get the most out of these, you'll probably want to use a microcontroller to decode the row-and-column combos. (We won't talk about microcontrollers in this book, but we still feel it was worth a mention.)

Piano Keyboards

This is an interface specifically designed for music. In the early days of electronic synthesizers, keyboard interfaces had an output called CV, or *control voltage*. Pressing a key resulted in the production of a particular voltage, which could then be interpreted by the synthesizer. A higher voltage means a higher frequency. If you can get your hands on a CV-capable keyboard, you can certainly use it to control whatever is on your breadboard. Digital keyboards, which send out instructions in the MIDI language, won't translate to your breadboard unless, of course, you invest in a swanky MIDI-to-CV converter.

Slide Switches

These are your garden-variety on/off switches—move the lever into a new position, and it'll stick there until it's reactuated.

Toggle (or Rocker) Switches

Like slide switches, toggles, also called rockers, can be pushed one way or the other. You can find toggles with different pole/throw combinations (see the bottom of opposite page), and they look super professional. You'll sometimes find toggles with LEDs inside, which light up all pretty when the switch is in a particular state.

DIP Switches

DIP stands for "dual inline package," which means the legs on it are perfectly sized for breadboard use. These switches have many little levers you can flick with a pair of tweezers or fingers (if you have very skinny fingers). A DIP switch is mostly used for prototyping or letting a consumer manipulate rarely changed settings. For example, old garage-door openers sometimes have hidden DIP switches that allow you to change how it interacts with the remote. It's far easier for a consumer to flip a switch than it is to rewire a circuit, especially if it's a set-it-and-forget it feature.

Rotary Switches

These switches are what you might think of as a dial. Spin the circular part around, and you'll spin through several switchable paths.

Pull Chains

These latching switches have a nice little chain on them, like what you'll find on a bedside lamp.

Up to this point, we've discussed only switches that make or break a circuit. But a switch doesn't have to exist in an on/off binary state. In fact, you can find switches that connect any number of things to any number of other things.

Switches connect *poles* to *throws*. If a switch controls just a single pathway—either broken or unbroken—it's referred to as a single-pole, single-throw (SPST) switch. If a switch allows a signal to flow to one of two outputs and the switch selects the chosen output, we call that a single-pole, dual-throw (SPDT) switch. Here are some examples of poles and throws.

SPST

A single-pole, single-throw switch is the simplest, with only one output and one input. Almost every on/off switch you find is an SPST—all it does is disconnect the main circuit from the battery.

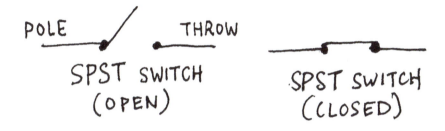

Adobe Stock-Siwapot Narukietmont

Railway-junction interchange point for train routes

SPDT

A single-pole, dual-throw switch is a bit more complex—instead of a drawbridge, it resembles a railway interchange, guiding trains from one track onto one of two additional tracks.

With an SPDT switch, we can have electrons flow in one of two pathways. This switch could be used to change the speaker a sound signal is coming out of, or, conversely, it could change the audio source being sent to your speaker. SPDT switches have three terminals, which can be wired as one input and two outputs or two outputs and one input.

The image below and on the next page show schematic symbols for different switches.

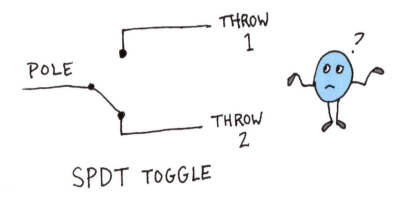

DPST

Dual-pole, single-throw (DPST) switches are just like the SPST drawbridges discussed above, except that they're like *two* drawbridges that can be raised or lowered simultaneously. This is helpful if you want one flick of a switch to effectively act as *two* switches:

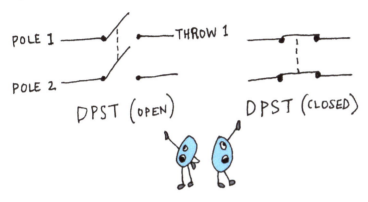

DPDT

A dual-pole, dual-throw (DPDT) switch is, appropriately, a little bit more complex. Instead of acting like two simultaneous drawbridges, a DPDT functions like two simultaneous railway interchanges: You can make all your trains take a path to the left, or you can make all your trains take a path to the right. There is no other option.

   I'll be frank with you: DPDT switches are rare for DIY music projects but are still common enough that they merit a mention in this chapter. At home, you'll find DPDT switches are what connects your wiring to your city power—the two poles are used for turning on the live and neutral connections together. Connecting the live and neutral connections at different times could have disastrous consequences.

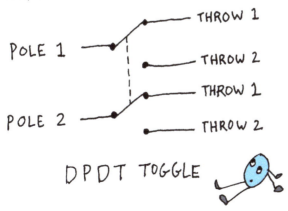

WHAT ABOUT ICS?

On a schematic, ICs are the least interestingly depicted. More often than not, you'll draw a rectangle or a triangle, label it with the IC's name, and draw the connections wherever they're cleanest. Then you'll write the numbers of the pins next to the connections.

Here is an image of a schematic for our basic LM386 power amplifier. Notice how the IC in the schematic doesn't have pins in the same places as our actual LM386 on a breadboard:

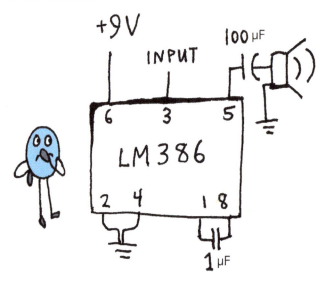

Notice, too, how all the important connections are there. Let's peruse the schematic and see what we can learn:

- Pin 6 of the LM386 connects to 9 V, while pins 2 and 4 connect to ground.
- There is a 1-microfarad capacitor between pins 1 and 8. Because the capacitor symbol is not polarized, it does not matter which direction it faces.
- Pin 3 is our input—this is where our audio signal will go.
- Pin 5 is touching the positive leg of a 100-microfarad capacitor. The negative leg of that capacitor connects to a speaker terminal. The other speaker terminal connects to ground.

And that's it! Isn't that simple?

It's important to remember that the schematic drawing of the IC probably won't have the pins in the same orientation as they'll be in real life. In fact, getting used to the fact that a schematic doesn't look like a breadboard is easily the most challenging part of this skill.

THE ORGANIZATION

Indulge me for a minute, and take a look at a map of the Rotterdam metro. It's a truly glorious map to behold—clean, geometric, and colorful. Little dots show us the stations, and the lines show us which stations connect to each other. It's a real thing of beauty.

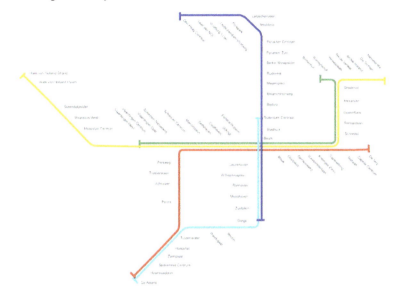

   This map has been around since 1968 and has become a beloved Dutch symbol up there with building windmills and subjugating other lands to build windmills. So it is with a heavy heart that I must tell you the Rotterdam metro doesn't look anything like the above picture. In fact, it looks much more like this:

Jules Blom

Yikes! This second map, a far cry from the clean lines of the first, reflects the actual geography of Rotterdam. It sacrifices legibility for geographic accuracy.

Rotterdam's metro map—and, truthfully, *every* transit map—is a bald-faced lie. It ignores the actual location of things and shows a grossly distorted version of reality. But this is a good thing! Imagine having to navigate around a city using the *true* map. It's hard to see at a glance where certain lines go, it's bunched up in some places and spread out in others, and there's barely enough space to make the writing big enough to be legible. For the everyday commuter, these distortions of reality don't particularly matter.

I'm bringing this up, dear reader, because metro maps are schematics. They show us a simple, easy-to-read abstraction of what our breadboard will look like. A schematic shows you what connects to what, but it does not show you where anything goes.

For example, there are many ways we could attach a resistor and an LED to a battery to light it up.

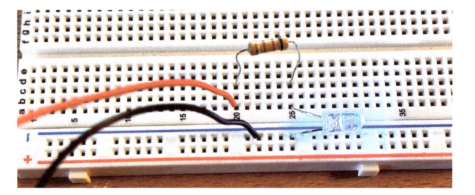

Or like this:

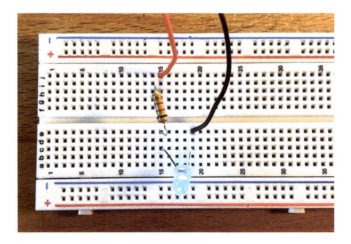

Or, heck, it's not particularly efficient, but this works too:

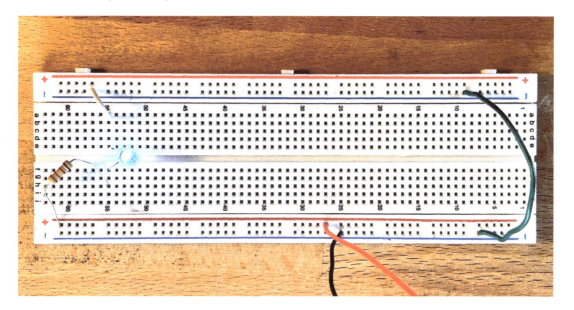

Notice how in each breadboard layout, although the parts are in different places, the basic schematic is maintained. When you build a circuit out, you get to choose where you'll put the parts! The schematic doesn't really care at all. All the schematic has to do is give you a circuit that works. And as long as the right parts connect to the right parts, your circuit will work.

Learning the symbols for a schematic diagram is 90 percent of the work, but the hard part to wrap your head around is that your breadboard will look nothing like your schematic.

Let's take a look at our basic oscillator circuit. In this schematic, we've gone one step further and provided you with an image not of the 4093 IC but of the individual NAND gate inside:

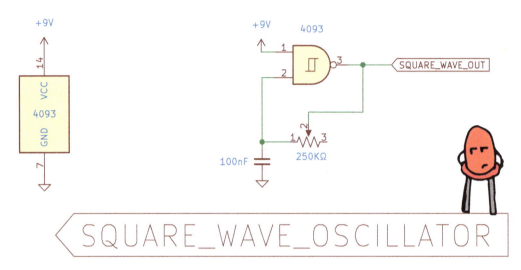

A schematic for our basic CD4093 oscillator, using pins 1, 2, and 3

Let's read this schematic and see what it can tell us:
- In the lower left-hand corner, we see the CD4093. This tells us to attach pin 14 to +9 V and pin 7 to ground.
- Now, we take a look at the NAND gate—notice that the three pins we need to work with are labeled 1, 2, and 3.
- Pin 1 connects to 9 V.
- Pin 2 touches one leg of a capacitor, and the other capacitor leg touches ground.
- Pin 3 is touching the nose (middle pin) of a potentiometer. One of the cheeks (side pin) is touching pin 2.
- Our audio output comes from pin 3, where we can send it to our LM386 or a consumer-power amplifier.

You'll notice that reading schematics becomes much easier the more you do it. Just break the problem into individual steps and conquer it that way.

For a slightly more complex schematic, let's look at our cricket circuit from the previous chapter.

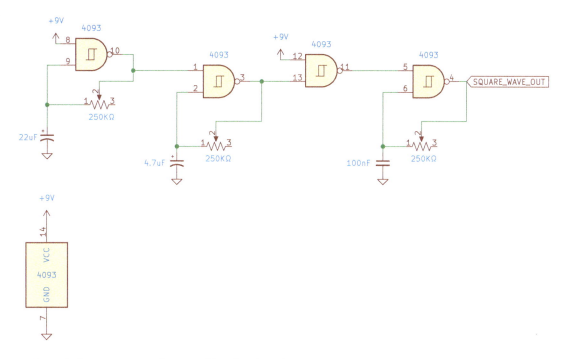

A Schematic for our CD4093 Electro-Cricket (see last chapter)

This schematic, while a little hairier, is still pretty simple to parse. Let's give it a shot:

- First, let's set up pins 8, 9, and 10 to oscillate. We can do this by attaching pin 8 to +9 V, a capacitor between pin 9 and ground, and a potentiometer between pins 10 and 9.
- Our output from pin 10 will attach to pin 1. Add a capacitor between pin 2 and ground and a potentiometer between pins 2 and 3.
- Our output from pin 3 will go directly into pin 13. Attach pin 12 to +9 V so the NAND gate is set up as an inverter.
- Finally, send our output from pin 10 to pin 5. Add a capacitor between pins 6 and ground and a potentiometer between pins 4 and 6.
- Send the output from pin 4 to a power amplifier.

Just like that, we've translated a schematic into a real, breadboardable circuit.
Schematics are a little bit frustrating, but we have found that after the initial shock, students come to appreciate them. You'll start to realize in the next chapter how big of a help it is to have a schematic. However, there is an easy way

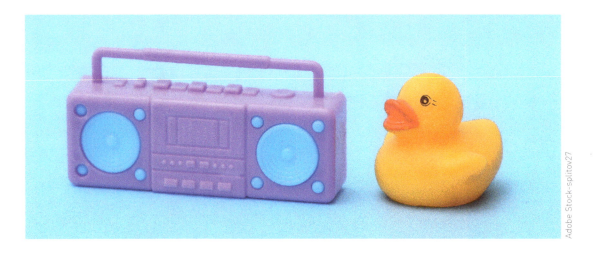

Adobe Stock-splitov27

to get quickly acquainted, and I promise we are serious about this.

Get a rubber duck.

For a few decades, computer programmers have sworn by a silly custom called duck debugging: put a rubber duck on your desk and, as you read a schematic, explain it slowly and clearly to your duck. You'll be surprised how simple schematics are when you present them in bite-size chunks.

Some final points to make about schematics:

• Sometimes, if you're reading a schematic from a different country, you might see symbols that are *ever so slightly different* from the ones we've printed here. For example, sometimes ground looks like a little triangle instead of the three parallel lines we use here. Most of the time, however, these symbols are fairly easy to interpret. Just remember that there is no universally accepted electronic schematic standard.

• You might see the terms *VCC*, *VDD*, *VSS*, and *VEE* on some schematics. *VXX* and *VDD* represent the *positive electrode* on your battery, and *VSS* and *VEE* represent the *negative electrode*, also known as ground.

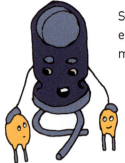

Schematics are a frustrating but necessary step to becoming an independent electronic-music tinkerer. And next time you get on a subway, be thankful your map is lying to you. 🎵

Handy Formulas

VOLTAGE DIVISION:

$$V_{OUT} = V_{IN} \left[\frac{R_2}{R_1 + R_2} \right]$$

↑ DIVIDED OUTPUT VOLTAGE

↑ INPUT VOLTAGE

↑ THE TWO RESISTOR VALUES IN YOUR DIVIDER (IN OHMS)

OHM'S LAW: $V = I \times R$

VOLTAGE (IN VOLTS) ↗

↑ CURRENT (IN AMPS)

↑ RESISTANCE (IN OHMS)

PASSIVE FILTER CUTOFF:

$$F_c = \frac{1}{2\pi RC}$$

CAPACITOR VALUE (IN FARADS)

↑ FILTER CUTOFF IN H_2

↖ RESISTOR VALUE (IN OHMS)

VOLTAGE GAIN:
(FOR INVERTING OF AMP)

$$Gain = \frac{V_{OUT}}{V_{IN}} = \frac{-R_F}{R_I}$$

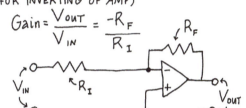

Common Interval Ratios

RATIO	NAME	NOTATION EXAMPLE
2:1	Octave	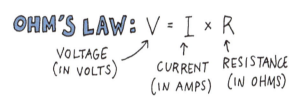
3:2	Perfect Fifth	
4:3	Perfect Fourth	
5:4	Major Third	
6:5	Minor Third	
45:32	Tritone	

Part
III

BIGGER CIRCUITS

Congratulations, you budding electrical engineer, you. You've made it quite far! You know your resistors from your capacitors, how to solder, and how to look at a schematic without getting anxious. You've reached a point where you can, honestly, build just about anything. If you have the parts, can read the map, and know how to put them together, you can happily put this book down and enjoy an extensive career as a circuit builder. Given the number of schematics available online, all you have to do is find the circuit of your dreams.

The remainder of this tome is filled with schematics for all sorts of increasingly complicated projects. We'll still walk you through how everything works, but we will slowly remove the training wheels as time goes on. We'll forgo step-by-step instructions in favor of schematics, and our images of breadboards will become increasingly scant.

For every project that remains in this book, we expect you already have these four things at the ready:

- **A breadboard**
- **A bunch of cables (jumper wires)**
- **An audio source**
- **An amplifier**

The projects that follow in Part III are our favorites—as musically adventurous as they are scientifically intriguing. We hope your imagination runs wild.

When I was a kid, I was taught that the thickest, deepest-pitched string on a guitar was the E string, and when you plucked it, you would hear an E2. This is a lie. Or, at best, it's a half-truth.

In fact, when you pluck an E string, you are hearing an E2, but you're also hearing a B3, a G#5, and literally dozens of other notes (albeit much more quietly). This is because different parts of the guitar are vibrating in different ways, and making different shapes. Imagine sitting on a roller coaster and noting how every rider jostles in their own slightly unique way. A rider's body and posture will determine *how* they jostle, as will their position within the roller-coaster car.

Similarly, different parts of the guitar vibrate at different rates and in different patterns. As the E string moves, it pushes and pulls on the bridge. The bridge transduces the sound wave into the guitar body, pushing and pulling on the big cavity of air inside the instrument. Some frequencies of sound will have a faster time finding the sound hole and escaping than others.

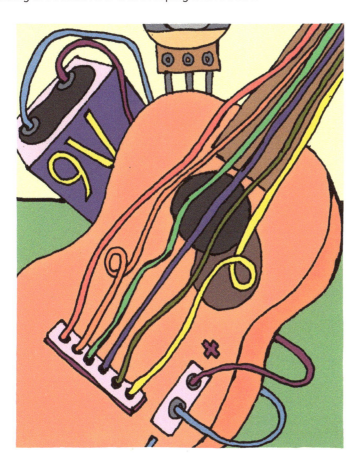

How to talk about pitch

We measure frequencies in hertz, each unit of which is equal to one wave cycle per second. We can think about wave cycles as the amount of time between "pops" of our speaker, or the distance from one square wave's peak to the next. If our speaker pops out 500 times a second, we are making a 500 Hz wave. Simple!

Musical pitch is different from frequency. While *frequency* has a rigorous, mathematical definition, *pitch* is an amorphous, human-defined term. We can convert these hertz values into musical notes by rounding them to the nearest note on the chart below. However, 500 Hz happens to not really coincide with any defined note—in our Western tuning system,[1] the closest we can round it to is the note B4. The *B* refers to the musical note, and the *4* refers to the octave. Are these letters and numbers arbitrary? You bet.

	Octave 0	Octave 1	Octave 2	Octave 3	Octave 4	Octave 5	Octave 6	Octave 7	Octave 8
C	16.35 Hz	32.70 Hz	65.41 Hz	130.81 Hz	261.63 Hz	523.25 Hz	1046.50 Hz	2093.00 Hz	4186.01 Hz
C#	17.32 Hz	34.65 Hz	69.30 Hz	138.59 Hz	277.18 Hz	554.37 Hz	1108.73 Hz	2217.46 Hz	4434.92 Hz
D	18.35 Hz	36.71 Hz	73.42 Hz	146.83 Hz	293.66 Hz	587.33 Hz	1174.66 Hz	2349.32 Hz	4698.63 Hz
Eb	19.45 Hz	38.89 Hz	77.78 Hz	155.56 Hz	311.13 Hz	622.25 Hz	1244.51 Hz	2489.02 Hz	4978.03 Hz
E	20.60 Hz	41.20 Hz	82.41 Hz	164.81 Hz	329.63 Hz	659.25 Hz	1318.51 Hz	2637.02 Hz	5274.04 Hz
F	21.83 Hz	43.65 Hz	87.31 Hz	174.61 Hz	349.23 Hz	698.46 Hz	1396.91 Hz	2793.83 Hz	5587.65 Hz
F#	23.12 Hz	46.25 Hz	92.50 Hz	185.00 Hz	369.99 Hz	739.99 Hz	1479.98 Hz	2959.96 Hz	5919.91 Hz
G	24.50 Hz	49.00 Hz	98.00 Hz	196.00 Hz	392.00 Hz	783.99 Hz	1567.98 Hz	3135.96 Hz	6271.93 Hz
Ab	25.96 Hz	51.91 Hz	103.83 Hz	207.65 Hz	415.30 Hz	830.61 Hz	1661.22 Hz	3322.44 Hz	6644.88 Hz
A	27.50 Hz	55.00 Hz	110.00 Hz	220.00 Hz	440.00 Hz	880.00 Hz	1760.00 Hz	3520.00 Hz	7040.00 Hz
Bb	29.14 Hz	58.27 Hz	116.54 Hz	233.08 Hz	466.16 Hz	932.33 Hz	1864.66 Hz	3729.31 Hz	7458.62 Hz
B	30.87 Hz	61.74 Hz	123.47 Hz	246.94 Hz	493.88 Hz	987.77 Hz	1975.53 Hz	3951.07 Hz	7902.13 Hz

1. This system is called the twelve-tone equal temperament.

THE HARMONIC SERIES

All these nuanced vibrations are what make a guitar sound like a guitar. The volume of the body, the ease at which the instrument vibrates, and the material the strings are made of all contribute to the instrument's tone color. A guitar with a wooden body and nylon strings will sound very different from an aluminum guitar with steel strings. Both of these guitars will sound E when you pluck the top string, but their tone colors are entirely different.

We can use a tool called a spectrum analyzer to look at a guitar plucking an E.

This image is called a spectrogram, and it shows us the different vibrations present in the sound. The yellow horizontal lines represent the different speeds of vibration occurring within the guitar. The big yellow line at the bottom is the loudest (and lowest) vibration—around 82 Hz, which is indeed the pitch humans have named E2. This lowest frequency is called the *fundamental,* as it's the loudest and the one our brains are most likely to latch onto. However, you'll also notice a bunch of other peaks, representing higher notes that are also present in the sound—the other, less-prominent vibrations. These higher notes are called the *overtones* of the guitar. Together, the fundamental and the overtones are referred to as *partials*.

Why is this so important? Because every sound's overtone series is unique. It's what makes a piano sound so different from, say, a violin or a xylophone *even when they are trying to imitate each other*. The pattern made by the spacings of partials is called the sound's *harmonic series*. I like to think of a sound's harmonic series as its sonic fingerprint. If we change the overtone series, we'll change the identity of the guitar. A sound's harmonic spectrum is called its *color*, or, if you want to sound like a musician, its *timbre* (pronounced "tam-burr"). You'll hear people refer to sounds with a wide spectrum of partials as *harmonically rich* and sounds with a limited set of partials as *harmonically poor*.

Here is a spectrogram of a saxophone playing the same note—E2:

And while we're at it, here's a CD4093 square-wave oscillator playing an E2:

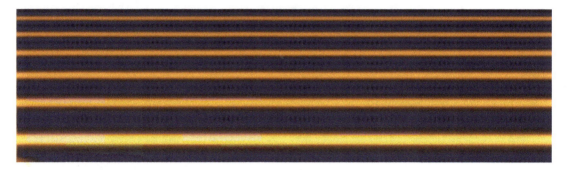

Interestingly enough, only one wave has no overtones—the humble *sine wave*:

A lot of people seem to think a sine wave is "just any curvy wave, but that really couldn't be further from the truth. A sine wave is a very special shape that exhibits a constant rate of change (like a weight on a spring). This is called *simple harmonic motion*, as it exists only as a fundamental—the only timbre in which playing an E results in literally just an E. A sine wave isn't "harmonically poor" so much as it is simply a single, isolated harmonic.

This chapter is all about timbre modulation—how to take a sound, change its overtone series, and play it back.

ADDITIVE AND SUBTRACTIVE SYNTHESIS

By this point, you've got a pretty good handle on the idea that synthesizers make sounds by drawing shapes with voltage. Different shapes produce different sounds, like how a triangle wave sounds different from a square wave. Up until now, we've been making sounds by mixing different waves together. Add a square wave to a triangle wave, and you'll get a whole new wave that looks something like the illustration below.

Synthesis means "making a new thing out of other things," and in this case, we've created a third wave by combining two others. This type of sound synthesis is called, creatively, additive synthesis. It's pretty simple but can be spookily effective.

In the 1820s, a scientist named Joseph Fourier noted a little mathematical quirk while studying how heat spreads through a system. He noticed that any

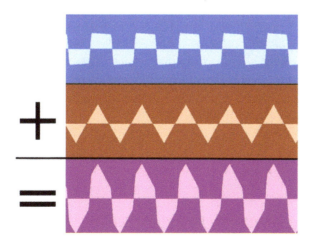

mathematical function (or, if you prefer to think of it as such, any *wave shape*) could be described as the sum of a bunch of sine waves. After all, sine waves have no overtones—just pick the frequencies you need and throw them together. This really does sound like sorcery, but it's totally true. And he figured this out in the *1820s*.

Fourier, of course, wasn't thinking about sound when he made this mathematical discovery, but it proved to be a keystone in the understanding of electronic music some seventy years later. In short, if you want to conjure the sound of a trumpet, or a banjo, or Ethel Merman singing "Baby Shark," all you have to do is combine several specific sine waves. Thaddeus Cahill's Telharmonium, mentioned way back in the first chapter, created sounds through additive synthesis.

As with all great things, there's a catch, and additive synthesis is no exception. While combining sine waves is *technically* all you need to make any timbre in the world, it sure as hell isn't efficient. Mimicking any arbitrary sound might need *hundreds* of oscillators to get it right. Couldn't there be an easier way?

In fact, there is. If additive synthesis is like making a collage, *subtractive synthesis* is like chiseling a sculpture out of marble. In the former, we make a harmonically rich wave by combining harmonically simple waves. In the latter, we *start* with a harmonically rich wave and *filter out* certain partials to make another wave. A filter, like the one in your coffee maker or your swimming pool, is a device intended to separate stuff from other stuff. (When something has no filter, like your weird conservative uncle, absolutely every bad thought that comes to his mind makes it out of his mouth—much to your mortification as he shares it with everyone else waiting in line on Splash Mountain.)

An audio filter is simply a little circuit designed to "pull out" a range of frequencies from the harmonic spectrum of your wave. As you can imagine, you can make a lot of harmonically complex waves with just a few oscillators and a filter—far more efficient than combining a zillion sine waves.

Subtractive synthesis, due to its comparatively high bang-for-your-buck marketing strategy, became the dominant form of sound synthesis in the 1910s and remains so to this day. It's certainly not the only type of synth you'll come across, but it's surely the most common.

Ring Modulation

A Dalek from Dr. Who, clearly threatening the poor photographer.

OK—so, you're saying to yourself, "When we add two waves together to get a new one that's *additive,* when we remove frequencies, it's *subtractive.* So, has anyone tried to multiply the two waves together? Is that even a thing?"

As it turns out, yes—and it is! There's a famous circuit called a *ring modulator* that effectively does just that. Multiplying two waves together has a very odd and enigmatic effect. It's the effect used for the Daleks in *Dr. Who* and on Ozzy Osbourne's voice at the beginning of Black Sabbath's "Iron Man." A low-frequency oscillator ring modulated with Ozzy's dulcet tones results in a sound reminiscent of singing behind a cooling fan.

It's called a ring modulator, by the way, because the diodes used in the circuit form a little ring. Electronics nerds might note that a ring-modulator circuit looks suspiciously like a full-bridge rectifier circuit with the diodes facing backwards. We'll show you how to build a version of this circuit in our "Modulation" chapter.

PROJECT: PASSIVE FILTERS

We can build an audio filter to exclude certain frequencies from the harmonic spectrum of any arbitrary sound. There are two basic types of audio filters—low pass and high pass. A *low-pass filter* removes all frequencies *lower* than a chosen cutoff frequency. A *high-pass filter* removes all frequencies *above* your chosen cutoff. Here, *passive* means that these circuits require no power. If you've ever played with a tone knob on an electric guitar, you've played with passive filters before. They're cheap, easy to make, and pretty interesting to think about (if you're the kind of person who likes thinking about things).

THE PASSIVE LOW-PASS FILTER

It's 2 a.m., you had a horrible day at work, and your neighbor is blasting Limp Bizkit through the wall. A single tear comes to your eye. But then you make an interesting discovery—through the wall, you're able to hear only the low rumbles of the bass and drums. Fred Durst's rage-inducing voice is totally muffled. Why is that?

It turns out that your wall, however thin, is acting as a low-pass filter, which lets low frequencies pass through while cutting out the high frequencies. Your neighbor's music, while poor in taste, is rich in overtones. The high-pitched sounds vibrate quickly and are relatively short, and literally get trapped by your walls. The thicker the walls, the more of the high end will be blocked off. The low-pitched sounds, which are much "larger," sail right on through as if that drywall wasn't there. To simulate the sound of a low-pass filter descending on an innocent square wave, make the following sound: "*eeeeeeee-wwwwwww.*" As your mouth closes, you'll hear the sound turn from bright to dark.

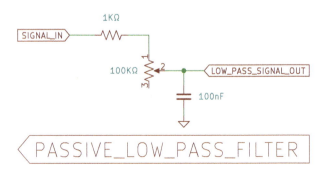

THE THEORY

You can make the world's easiest low-pass filter with just two passive components—a resistor and a capacitor (Figure **A**).

With just these two parts, we can calculate this filter's cutoff frequency:

$$Fc = 1/(2\pi*R*C)$$

Fc is the cutoff frequency in hertz, *R* is our resistor value in ohms, and *C* is our capacitor value in farads.

With the above example, we're using a 10 kilo-ohm resistor—that's 10,000 ohms. We're also using a 100 nF capacitor. There are 1 billion nanofarads in a farad, and with a little algebraic wiggling, we can say that a 100 nF capacitor has 0.00000010 farads.

$$Fc = 1 / (2\pi*10,000*0.00000010)$$
$$Fc = 1/ (0.0062831853)$$
$$Fc = 159.15$$

This means our filter—theoretically—should remove all partials higher than 159.15 Hz from our signal (letting the frequencies *lower* than 159.15 Hz pass through). Filters with fixed-value resistors and capacitors are great for knocking out unwanted hums. For instance, you could tune a filter to remove the 60 Hz electrical-mains hum you hear in North America (or the 50 Hz hum you hear everywhere outside North America).

Of course, for musical purposes, it's a lot more fun to have a filter you can twiddle around. Replace the resistor with the cheek and nose of a potentiometer, and you now have a *variable low-pass filter*.

INSTRUCTIONS:

1. Connect the output of your square wave to the cheek of a potentiometer. Connect a 100 nf (0.1 µf) capacitor between the potentiometer's nose and ground—remembering that the smaller leg always connects to ground (Figure **B**).

B

2. To hear the filtered output, gator-clip that wire to the tip of an audio cable, then gator-clip the sleeve of the audio cable to breadboard ground. Listen to the output while fiddling with the potentiometer's resistance. You should hear a subtle but unmistakable filter sweep!

Why Does This Happen?

A filter works because of something called *reactance*.

We've already talked about resistance, which is how much any given component will push back on the flow of current. Reactance is resistance applied to frequency—some components push back at some frequencies more than others.

To best illustrate why a filter does what it does, it's helpful to remember what a capacitor really is: two conductors physically separated by a layer of insulator. This allows positive charges to gather on one conductor, negative charges to gather on the other, and the two opposite charges to pull each other together. By pulling on each other, the capacitor can squeeze more charge into a small space than a normal wire could afford. However, the individual electrons never jump the gap between the two conductors—it's like a rubber membrane between the two. Charges on one side might be able to "feel" charges on the other side of the membrane, but they don't actually penetrate it.

Imagine rhythmically knocking on one side of a rubber curtain. Fast-moving wiggles will distort the rubber and cause the charges on the opposite side to wiggle. Slowly accumulating charges won't distort the rubber nearly as effectively, and those much more nuanced changes will go unnoticed.

A slightly fancier way of thinking about this is to consider that a high-pass filter allows AC to pass through unimpeded but blocks DC signals. A low-frequency signal moves a lot slower than a high-frequency signal, so you can think of a low pitch as having more "DC-ness" to it than a high pitch.

In a low-pass filter, the higher frequencies escape the signal by making it to ground through the capacitor. Everything left behind—the lower frequencies—sails out of our output.

THE PASSIVE HIGH-PASS FILTER

A low-pass filter lets frequencies *below* a cutoff *pass* through. A high-pass filter lets the frequencies *above* the cutoff *pass* through. Really creative naming here, guys.

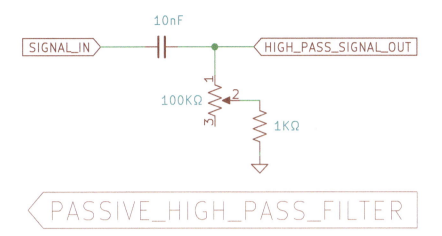

PASSIVE_HIGH_PASS_FILTER

INSTRUCTIONS:

A high-pass filter works much like a low-pass filter, except that the resistor and the capacitor are reversed. You can calculate the cutoff using the same equation we used for the low-pass filter, but of course, it's more fun to build one with a potentiometer:

Why Does This Happen?

A high-pass filter pushes the lower frequencies through a resistor to ground. By reducing the voltage of our input signal, we'll change what speed our capacitor starts "reacting."

A slightly more advanced guide to potentiometer wiring

Up until now, most of our potentiometer usage has been pretty *laissez-faire*. However, if you pay slightly more attention to which pins are in use, it becomes easier to build user interfaces that work intuitively.

As we discussed way back in Part I, a potentiometer is merely a big runway of graphite strapped between the two cheek pins. The nose is attached to a wiper which can move up and down the runway, changing the resistance between the nose and either cheek. Because the cheeks are attached to each other, the resistance between the nose and one cheek will complement the resistance between the nose and the other cheek. The sum of both resistances will total the potentiometer's written resistance.

This means if your potentiometer is twisted all the way to the left, the resistance between the left cheek and the nose will be practically zero. If we were to attach this to our 4093 oscillator, we would notice the potentiometer produces a *higher* pitch when twisted to the left, and a *lower* pitch when twisted to the right. While this is totally fine from an electronics perspective, we humans typically like it when knobs *increase* in value as you move them clockwise. In order to engineer our interface like so, all we'll have to do is redirect the wire heading to the leftmost cheek to the rightmost cheek.

Many first-time synth builders try to accomplish this by swapping where the potentiometer wires go on a circuit (and not on the potentiometer). For example, a lot of people assume that switching the potentiometer wires on pin 2 and pin 3 of a 4093 oscillator will also switch the direction of increasing pitch. However, this is not so (try it out)! The reason is because the oscillator only cares about the resistance in between the two pins—which doesn't change based on polarity.

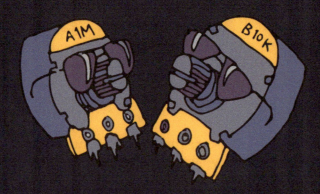

PROJECT: ACTIVE FILTERS

MATERIALS:
- 2 x 10 kΩ resistor
- 2 x 100 kΩ potentiometer
- 100 nF capacitor
- 10 nF capacitor
- TL074 IC

Passive filters get the job done, but because they don't have any powered components, they tend to remove a good bit of energy from our audio signals. As a result, passive filters can't make a signal any louder than it was before. However, if we get acquainted with a little device called an *operational amplifier*, or op amp, we can redesign our filters so they're significantly cleaner. These active filters work the same way the passive filters do, with the addition of an amplification stage.

The easiest active filter we can vouch for uses the TL074 op amp. It's a simple 14-pin integrated circuit that—like the 4093—has four subunits within. These subunits are not NAND gates, but rather *amplifiers*. Each low- or high-pass filter will use one op-amp stage. Therefore, we can squeeze four filters out of one chip (Figure **A**).

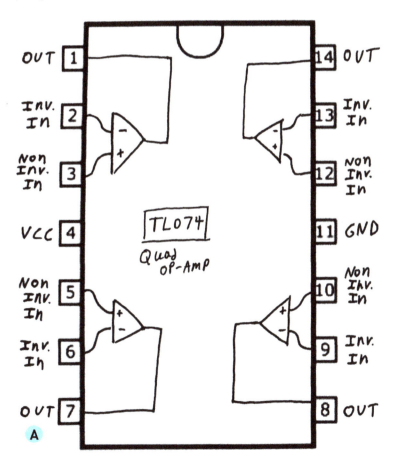

A

THE ACTIVE HIGH-PASS FILTER

Here is a schematic for an active high-pass filter. It's pretty simple! Remember that pin 4 is for your 9 V power and pin 11 is ground. Be careful to pay attention to where the inputs and outputs are—we can "listen" to the four op amps at pin 1, pin 7, pin 8, and pin 14 (Figure **B**).

Here's what it looks like on a breadboard (Figure **C**).

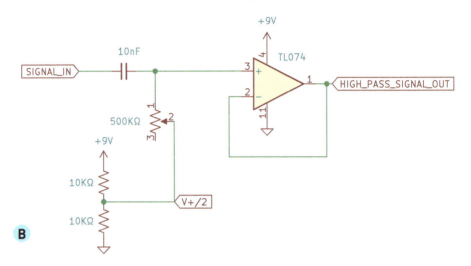

B

C

THE ACTIVE LOW-PASS FILTER

The active low-pass filter, no surprise, is exactly the same as the high-pass filter, but with the position of the capacitor and the potentiometer switched (Figure **D**).

Here's what it looks like on a breadboard (Figure **E**).

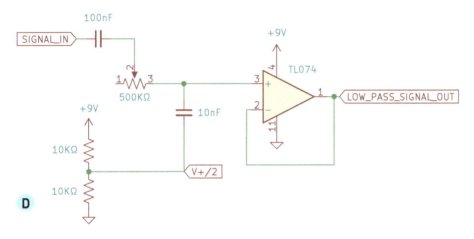

D

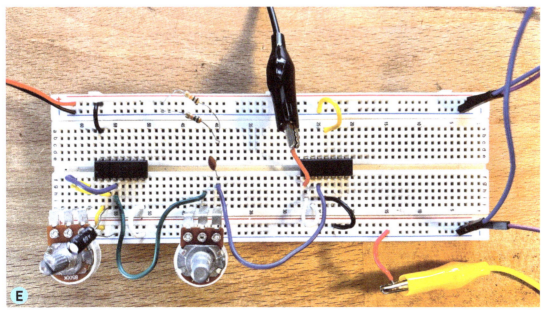

E

THE ACTIVE BAND-PASS FILTER

A *band-pass filter* lets a "band of frequencies" pass through—you can pick how wide or narrow the band is. For example, you could tune a band-pass filter (Figure **F**) to only allow a specific range of notes to pass through. A passive band-pass filter is as easy to build as a combination high- and low-pass filter (Figure **G**).

Use the two potentiometers to set your bandwidth. When the potentiometers have a greater difference in their resistance, you are letting a wide band of frequencies pass to your amplifier. When they have a small difference, you have a narrower band. When your lower-bound potentiometer is at a higher resistance than your upper-bound potentiometer, you'll have a *notch* filter—one that knocks out the frequency band instead of letting it pass through. 🎵

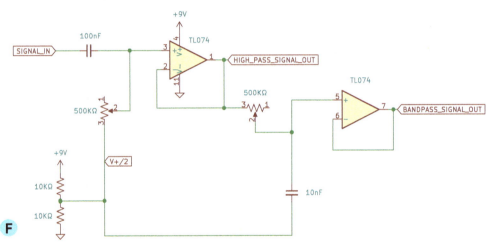

F

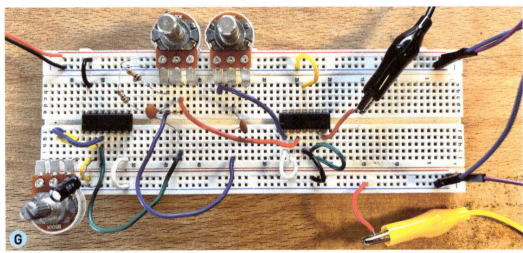

G

This chapter is all about creating circuits that can harmonize with your oscillator, hence the chapter name, "harmonization." Ironically, the chapter's theme is really "division." I'll explain.

In music, "harmony" is a fancy word for playing multiple notes at the same time. When you harmonize together at choir practice, you sing a variety of *different* notes that form a chord.

In math, "division" is a fancy word for splitting things up into equally-sized groups. When you divide a box of donuts between your choir members, you split the number of donuts between the singers as evenly as possible.

Division and harmonization are—from a technical standpoint—exactly the same process. Every week at barbershop quartet practice, you harmonize by listening to the frequency your friend is singing, transforming your frequency by a factor of X, and singing the new frequency.

Say, for example, we want to sing an octave higher than our friend. An octave is music theory speak for "a power of two of a given frequency." Sing double, quadruple, or octuple the frequency of your friend, and you'll sing the same note one, two, and three octaves higher. To pitch a note up an octave, all we have to do is figure out what frequency is being sung, double that number, and sing back the new pitch. 1000 Hz is exactly one octave above 500 Hz. Non-octave harmonies can be produced by dividing a frequency by numbers that aren't powers of two. For example, divide a frequency by 1.5 and you'll produce what musicians call a "perfect fifth." Of course, no singer is actually thinking about frequency division in their head—they've learned how to sing harmony heuristically, not algorithmically. However, this is a book about working with machines, which have no soul or heuristic capabilities. Division will have to do.

Harmony is a process that can also be thought of as multiplication. For example, multiplying any frequency by the fraction 4/3 (1.333...) will produce a perfect fourth higher than your initial note. However, multiplying numbers with electricity is a little complicated. *It's much easier to generate harmony by dividing frequencies by one another than by multiplying them.* (However, in Chapter 13 on phase-locked loops, we will show you how to create a frequency-multiplication circuit.)

This chapter is largely about two ICs—the 4040 and the 4017. Both are ICs that will help us divide frequencies, but both operate with slightly different rules. We'll start with the simpler of the two.

MATERIALS:
- 100KΩ potentiometer
- 100nF capacitor
- 4093 IC
- 4040 IC
- Amplifier with audio cable

PROJECT: OCTAVE HARMONIZER

INTRODUCING THE 4040

The CD4040 is a 16-pin *binary divider*, which means it splits things in two. If you recall, splitting a frequency in two is just math speak for harmonizing an octave below. Just like our other CD4000 series chips, the upper left-hand corner (pin 16) attaches to +9v, and the lower right-hand corner (pin 8) attaches to ground. Pin 11 is the "reset" pin, which must be attached to ground. Let's take a look at the chip in detail (Figure **A**).

Outside of the power pins, the 4040 has one input (pin 10) and twelve outputs (pins 1 to 7, 9, and 12 to 15). When the 4040 receives a square wave at pin 10, it will subdivide that frequency a number of times. Pin 9 (marked /2) will output a square wave with *exactly half the frequency* as the input. Pin 7 (marked /4) will output a square wave with *exactly one quarter of the input frequency*, and so on. This is to say, if we build a square wave oscillator, tune it to 420 Hz, and feed it into our CD4040, the outputs will instantly sync up to 210 Hz, 105 Hz, 52.5 Hz, and so on and so forth (Figure **B**).

Looking at the waves, it's pretty clear how the chip works. Our (/2) output simply waits until an even number of pulses have made it into our clock input. The (/4) output waits until an even number of pulses have made it into our (/2) input, and so forth. On every other clock cycle, each input nudges the next one down the line.

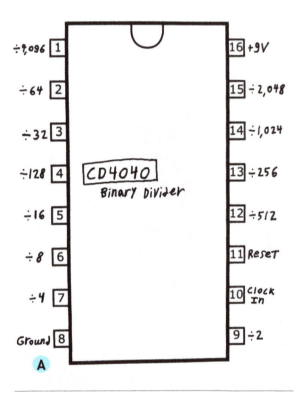

A

B

INSTRUCTIONS:

Take a look at this handy schematic—with one oscillator and one divider, we can cook up a smorgasbord of delicious octaves (Figure **C**).

Let's see what this lil' fella can do.

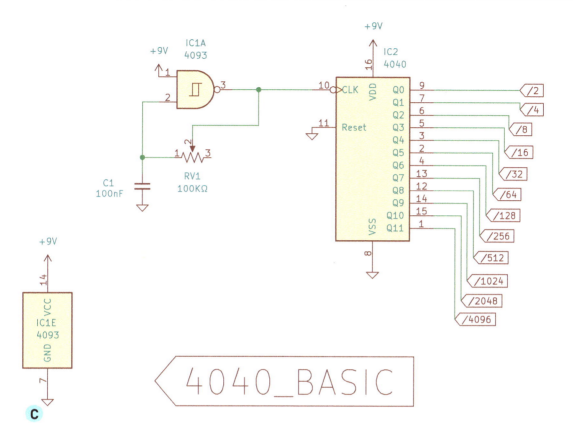

C

1. Build a square-wave oscillator with your CD4093.

2. Put a CD4040 nearby, and attach the power wires—pin 8 to ground, pin 16 to +9V (Figure **D**).

E

F

3. Pin 11 is our "reset" pin. When it is attached to +9V it acts like a pause button. We won't be using it for this demo, so attach it to ground (Figure **E**).

4. Pin 10 is our "clock input" pin. This is where our oscillator's output will go into the 4040—go ahead and attach a high-frequency square-wave source (Figure **F**). (If you attach a square wave that's too low frequency, you might not hear the outputs as "pitches," but instead as "clicks.")

5. With your amplifier, take a listen to some of the sub-octaves (I recommend listening to pins 9, 7, and 6 for the most obvious) (Figure **G**).

By dividing signals evenly in two, we've used a binary divider in a way it wasn't really intended—as an octave harmonizer.

A fun usage for these harmonies is to create some chiptune (kinda). Pick a few octaves that are your favorites and mix them together using resistors (Figure **H**). Instead of the kinda wimpy single-frequency square waves we've been playing with before, we can now make surprisingly rich synth voices by adding additional octave undertones.

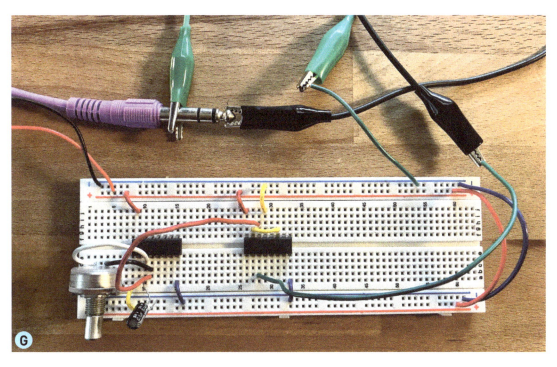

G

H

PROJECT: SUBHARMONIC GENERATOR

INTRODUCING THE 4017

The CD4017 is a "decade counter." A decade is what engineers call "a set of ten things," be they years, commandments, or "things I hate about you." Its job is to count from one to ten and then restart. Hookup is simple: as you might expect, the upper left hand corner (pin 16) attaches to +9v, and the lower right hand corner (pin 8) attaches to ground. It looks like what you see in Figure ⓘ.

The 4017 has one input (pin 14) and ten outputs (pins 1 to 7, and 9 to 12). The input is a square-wave clock. The ten outputs act like fingers—they're numbered one through ten, and as the IC counts it shoots current out of each finger like some Greek god. First, it will output current from "finger one" (pin 3, marked "step one") and only after the oscillator has ticked forward one count will it shoot electrons from "finger two" (pin 2, "step two"), and so forth, until the cows come home.

This is a fun chip if you want to make a lot of lights turn on one at a time. Send an oscillator into the clock, and an LED (with 330Ω current limiting resistor) between each output (long leg) and ground (short leg). Assuming your oscillator is ticking slowly enough, you will see the LEDs chase each other around the breadboard. We don't really know the utility for this, but we enjoy it nonetheless. Whee.

On the next page is a schematic of a CD4017 in its most basic operation, just aching for ten LEDs (Figure ⓙ).

Ok, that's enough epilepsy risk for today. What's the musical purpose of a chip like this? If we were to hook up an amplifier to 4017 pin 3, we would hear one click for every ten ticks of the oscillator. That means an input frequency of 1200 Hz (what musicians call a D6) would result in pin 3 outputting 120 Hz (what musicians call a

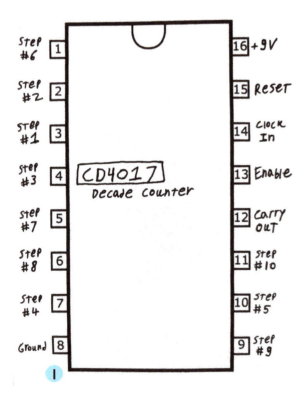

Step #6 | 1 — 16 | +9V
Step #2 | 2 — 15 | Reset
Step #1 | 3 — 14 | Clock In
CD4017 Decade Counter
Step #3 | 4 — 13 | Enable
Step #7 | 5 — 12 | Carry Out
Step #8 | 6 — 11 | Step #10
Step #4 | 7 — 10 | Step #5
Ground | 8 — 9 | Step #9

ⓘ

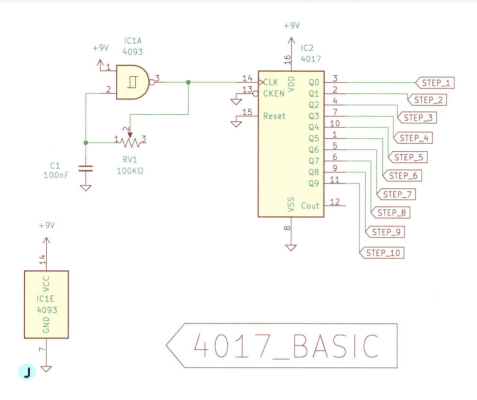

B2). Sensational, a chip that can divide by 10! If we were to listen to the oscillator *and* the 4017, we would hear a B2 and a D6 together (a minor third separated by three octaves). By sending that D6 into a 4040, we can transform it into a D2 (and get a minor third within one octave, or "closed position").

What makes the 4017 especially useful is that it doesn't *have* to count to ten. It can count to any number smaller than ten and reset itself. We can take advantage of this with the "reset pin," which is pin 15. Whenever pin 15 receives a burst of current, it automatically pushes the 4017 to the first step. So, for example, if we want our 4017 to count to 3 instead of 10, we will attach the pin that "lights up" on Step 3 to our reset pin. Then, our 4017 will count from 1, reset after step 4 is called, and start over again. Voilà—instead of dividing by 2, we're now dividing by 3.

INSTRUCTIONS:

1. The power wires, just like the 4040, obey a familiar shape. Pin 8 goes to ground, pin 16 gets attached to the +9v row.

2. Pin 13 is our "enable" pin. Attach it to ground. When pin 13 receives a high voltage, it will pause the counter (we don't need this feature yet).

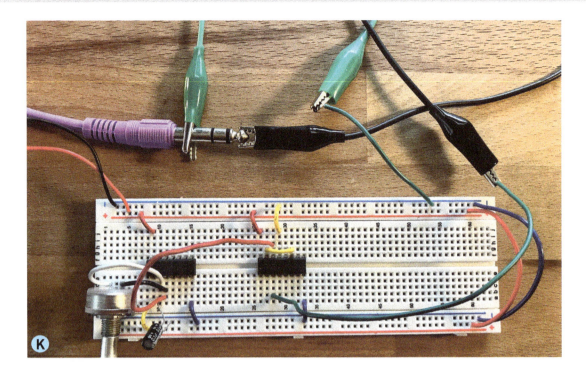

3. Pin 14 is our "clock input" pin. Attach a square wave.

4. Pin 15 is the "reset" pin. Attach it to any of the output pins on the 4017.

5. Pin 3 is the "first finger" it will count with. Plug this signal into your amplifier (Figure **K**). Take a listen, switch where the reset pin is attached to, and see if you hear a difference.

So what?

This chip is so much more interesting than it seems at first. In fact, there are two pretty cool musical things we can do with this—the first is *polyrhythmic generation*. My favorite analogy is to imagine two gears with a slightly different number of teeth. As the gears turn together, they won't "meet up" at their starting position until a particular number of rotations (the *least common multiple* of their respective numbers of teeth) is completed.

With a 4017 and a 4040 together, you can make all sorts of ratios that don't divide into each other evenly. For instance, with a two-step pattern on a 4040 mixed with a three-step pattern on a 4017, we can derive a rhythm called a *hemiola*.

The second usage is, as the chapter title says, harmonization. In theory,

harmonies are no different than polyrhythms. Harmonies are just rhythms pulsing so quickly that your brain interprets them as a musical interval. A binary divider can only derive octaves, but a decade counter can chop up frequencies ten ways! Music theory junkies are probably noticing what's coming out of their speakers right now—we've created an interval close to a major second, but with our two voices an octave apart. If we were to set up two 4017 chips in parallel and run the same clocking oscillator into both of them, we could arbitrarily set our reset pins to whatever ratio we would like. For instance, if we mixed the outputs of one 4017 set to count to 6 and another 4017 set to count to 5, we would derive a minor third.

You can derive your own scales by dividing frequencies into whole-number ratios. This tuning system is called *just intonation*, and it's perhaps the easiest tuning system to work with. The catch, however, is that most instruments you come into contact with aren't tuned justly—they instead obey the artificially smooth twelve-tone equal temperament system that we mentioned in the last chapter. The scales that are typically used in Western music don't obey the logical ratios of just intonation, and almost always have to be tuned by hand. That being said, the intervals are *pretty close*—but not exactly the same.

If you're interested in figuring out how you can harmonize by your-own-basic-interval-of-your-choosing, please consult this handy chart below:

Ratio	Name	Example
16:15	Minor 2nd	C3 and C#3
9:8	Major 2nd	C3 and D3
6:5	Minor 3rd	C3 and Eb3
5:4	Major 3rd	C3 and E3
4:3	Perfect 4th	C3 and F3
45:32	Tritone (Augmented 4th)	C3 and F#3
3:2	Perfect 5th	C3 and G3
8:5	Minor 6th	C3 and Ab3
5:3	Major 6th	C3 and A3
9:5	Minor 7th	C3 and Bb3
15:8	Major 7th	C3 and B3
2:1	Octave	C3 and C4

deriving weird intervals

Some of these intervals are quite a bit easier to set up than others. A perfect 4th would be simple: looking at the chart, we see a ratio of 4:3. You can think about it as multiplying by 4/3 (or, for that matter, dividing by 3/4). However, it's much easier to think about it as one clock ticking every four pulses, and another clock ticking every three pulses. Use the reset pin to make a 4017 a three-step counter, and then use another 4017 (or a 4040) to create a four-step counter. If you use resistors to mix the two outputs together, you'll hear a perfect 4th (in one octave

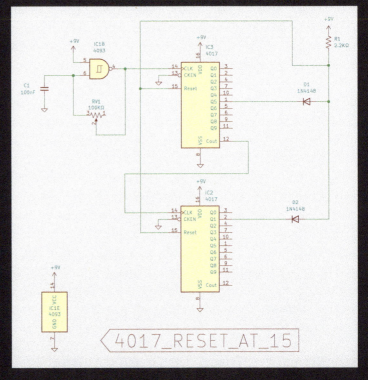

or another, if you're listening to a lower-octave output on the 4040).

A major 7th, on the other hand, is considerably stranger: a ratio of 15:8. Yikes! We can create a signal that outputs a pulse every 8 counts with either the 4017 (because 8 is less than 10) or the 4040 (8 is a power of 2). But how can we make a signal that pulses every 15 beats? Fortunately, the 4017 provides a means of *cascading*, which means we can chain up multiple chips to count to higher numbers. One 4017 can count up to 10, two can count up to 20, and so forth. The secret is pin 12, which acts as a number carrier. Once your first 4017 hits 10, pin 12 can be used to kick-start a second 4017.

To make a 15 step counter, set up two 4017s side by side. One will serve as our "tens place," the other as a "ones place." Connect pin 12 from the "tens place" IC to the enable pin of the "ones place." We will want to reset these two ICs after 15 counts—or when the tens place is outputting a 1, AND the ones place is outputting a 5. We can use two diodes to create an AND gate—a device that lets current pass through only when both ICs are displaying the right count. The output of our AND gate will be used to reset both 4017s when the number 15 pops up.

We will be talking much more about diode AND gates in future chapters.

PROJECT: 10-STAGE WAVETABLE GENERATOR

MATERIALS:
- 10 x 10KΩ resistors
- 10 x 10KΩ potentiometers
- 100KΩ potentiometer
- 100nF capacitor
- 4093 IC
- 4017 IC

A very easy and satisfying project uses one oscillator, one decade counter, and a whole bunch of potentiometers. By twisting the potentiometers, we can dial in a *tremendous* variety of timbres—from nasal and reedy to rich and bass-y. And this is *before* we add any filters to the mix.

A big trend in the 1980s was to make synthesizers that let you scroll through a "table" of different oscillator shapes. This circuit isn't exactly that, but through the years many DIY enthusiasts have come to call it a "wavetable synthesizer" because it sounds really flashy.

To explain how this circuit works, it's helpful to think about it moving incredibly slowly. In the schematic on the next page, all steps of the 4017 are being sent to our amplifier providing a pop-pop-pop sound as we sequence through each pin. By placing a potentiometer between each step and the amplifier, we can choose how loud each pop will be. With ten steps, we can make a variety of very boxy-looking shapes (Figure L).

These boxy-looking shapes, when played back at audio rate, all sound incredibly distinct from one another—truly! Twiddle your potentiometers around, and listen to how the timbre of your note changes. This is low-effort ambient music at its finest.

In the breadboard image of this circuit back on page 185, we are sending an oscillator (far left) into a 4017. On each of the 4017's outputs we've put a yellow wire to connect it to the nose of a potentiometer. Each of those potentiometers, in turn, has a cheek routed to an additional breadboard where the signals are mixed together through resistors and sent to the amplifier. You don't have to do this on multiple breadboards, of course, but it makes the image a little cleaner.

My favorite variation of this circuit is to replace the ten potentiometers with ten photoresistors and aim them in different directions. Play some flashlight tag with this circuit, and note how easy it is to change the timbre by swiping your hand around.

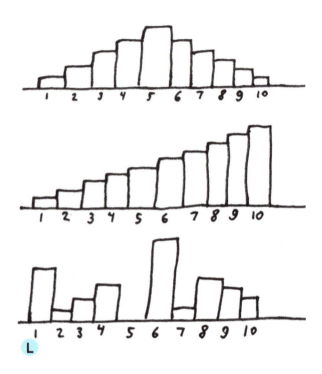

L

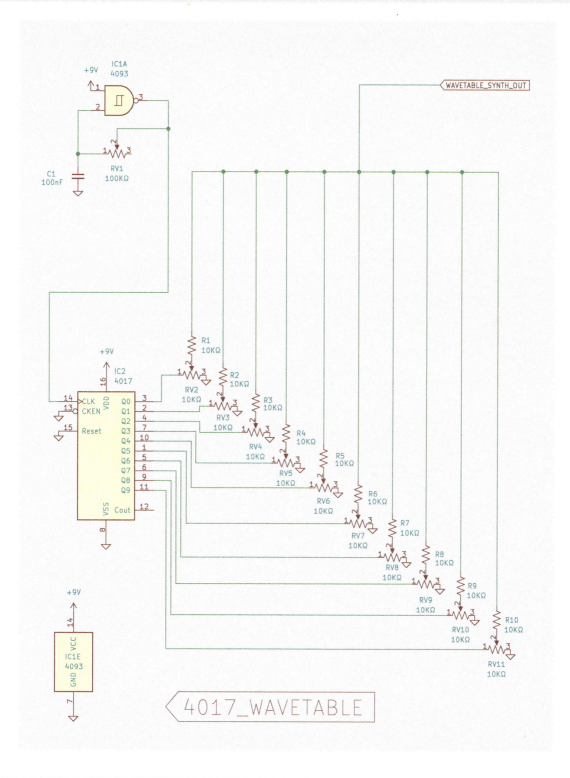

MATERIALS:
- 2 x 100KΩ potentiometer
- 100nF capacitor
- 10uF capacitor
- 4093 IC
- 4017 IC
- 4040 IC
- 4053 IC

PROJECT: THE FM YODELER

INTRODUCING THE 4053

What if—WHAT IF—we could have our harmony change intervals every once and a while? What if we could change the number we're dividing by every once in a blue moon? We can do this with the 4053 switch chip.

If we wanted to change the harmony of our oscillator, we could manually rewire the reset pin on the 4017, or the output that's being sent to our amplifier. A *switch chip* can do this for us. When a switch chip receives a signal from a square wave, it routes the current to another place.

Let's take a look at the 4053 now (Figure **M**).

You can see how the 4053 Switch Chip has three sets of SPDT switches. Like we discussed back in the Schematics chapter, an SPDT switch is like a railroad turnout—it can route one track to one of two places. When input A, B, or C receives a high voltage, it moves the arm of its respective switch. For example, pin 4 is normally connected to pin 5, but when a voltage is received at pin 9 it will flip and reconnect pin 4 to pin 3.

We only need one SPDT switch, and we're going to use it to change where the reset pin on the 4017 is routed to—thus changing the interval we're harmonizing with. Here's a schematic that will toggle between a division of 3 and a division of 2—a leap of a perfect 5th.

This schematic on the next page (Figure **N**) is reminiscent of how a person yodels. And who doesn't love a good yodel? Make a swooping sound from low to high with your voice, and notice how your voice seems to flip into a higher register just above your speaking pitch. This switch in vocal resonance areas is the dividing line between your "chest" and "head" voices, at least among choir enthusiasts.

Yodelers make their cliché "lay-ee-oo" sound by rapidly moving between their chest and head voices, creating a lovely if not grating alternation between two different wave shapes and frequency divisions. Just grab a 4053 and you'll be transported right back to Switcherland. ♪

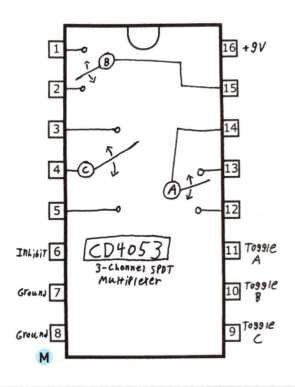

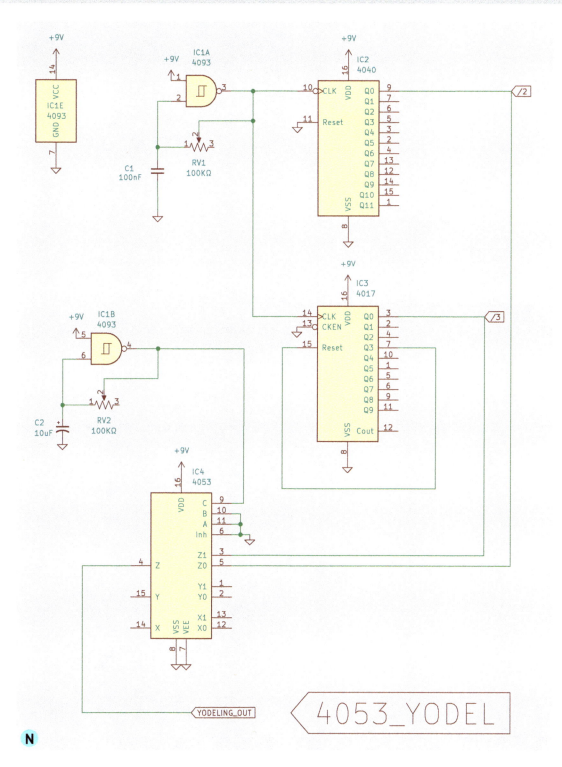

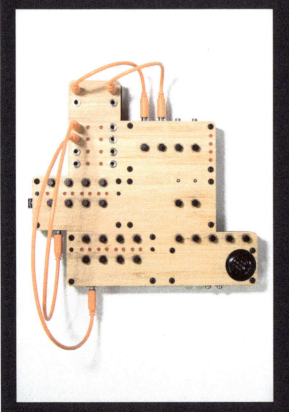

Musical Innovation: Modular Divider Box

Artist: Shane Culp

About: Shane is a musician and visual artist who builds a lot of instruments for musicians and his own group, Mission Giant. Shane's modular creations are beautifully minimal, and the odd placement of the jacks allows for patch cables to extend in all sorts of interesting directions. At the core of many of these creations are divider chips, used both as polyrhythmic clocks and as note harmonizers. A nice touch are the red indicator LEDs, which flash whenever its respective output peaks.

10

MODULATION

Modulation means a lot of things to musicians. If you change the key in the middle of a song (a common musical theater trick), you'll have *modulated* the key. If you're playing drums and suddenly shift to half-time— ta-da! You've performed a metric *modulation*. To make matters more confusing, there's a third musical meaning for this term that has absolutely nothing to do with the first two.

In electronics-ese, modulate means "change a property of a sound." Whenever you twist a potentiometer on a filter, you're providing a human modulation source. Back in our "Chaining Oscillators" chapter (see page 110), we used one square-wave oscillator to turn on and off a second, audio-rate oscillator—it sounded like a smoke detector. This is a basic form of modulation—one oscillator generates a *control signal* and the other produces the *audio signal*.

One of the beautiful things about electronic signals is that they can provide their own modulation sources. Instead of manually twisting a potentiometer up and down, we can build a circuit that'll do it for us. A modulation source is a little bit like having a third hand—it can change a parameter over and over again.

In this chapter, we're going to create some new shapes for control signals—far more interesting than the on/off square wave. We'll use these to modulate much more complex things such as filter cutoffs, pulse width, stereo panning, and volume.

Two components of a classical synthesizer exist *exclusively* for the purposes of modulation:

An *LFO*, or low-frequency oscillator, provides a very slow but predictable control voltage signal. It can be used to slowly modulate a parameter, such as the frequency of an oscillator or the cutoff of a filter. I think of an LFO like a ghost hand that can twiddle a knob back and forth so I can keep my hands free for nefarious purposes.

An *EG*, or envelope generator, provides a single burst of control voltage and the ability to morph the shape of that burst. Often, the EG is hooked up to the amplifier to provide a volume contour when a note is articulated. By changing the shape of the volume envelope, we can make an oscillator change from the sound of a plucked string to a bowed string. Of course, the EG doesn't only have to modulate the amplitude—it can just as easily be used to modulate a filter cutoff, or even an LFO's frequency.

The key similarity between an LFO and an EG is that both produce voltage signals you can't hear. That's because they're not audio—they're a control signal that's modulating another parameter. The key difference between an LFO and an EG is that one produces a repeating modulation signal and the other produces a single burst of signal.

Modulation is quite fun because it's a simple technique that can be used to make a really wide array of sounds. And once you add a modulation source to another modulation source, you'll realize just how unlimited your toolbox really is.

Let's begin!

MATERIALS:

- 2 x 250 kΩ potentiometers
- 2 x 100 nF capacitors
- 4.7 µF capacitor
- 47 µF capacitor
- LDR
- 330 Ω resistor
- LED
- 4093 IC

PROJECT: THE VANTASTIC VACTROL

Our simplest modulation source also happens to be a classic. Here's a little practical experiment for you. In Figure **A**, I've got an oscillator going into a high-pass filter.

Now, let's pull out the potentiometer and replace it with a photoresistor (Figure **B**).

Ta-da! An instant light-controlled filter. Take this into a dark room with a flashlight and try to "draw out" a few shapes with it. You'll hear a different effect if you jerk the flashlight around like a spooked cemetery caretaker versus if you swing it around gently like a graceful modern dancer. Yes, I know this sounds goofy, but this is more or less the gist of the following circuit.

Our very first breadboard circuit, a strobe light, will come in handy right about now. Let's rig up another oscillator to provide a slow flash to an LED. If we place the LED right next to our photoresistor, we'll notice it turns our filter cutoff from one level to another (Figure **C**).

A

B

C

When the LED flashes on, the filter cutoff changes. By manipulating the rate of the LED flash, we can make a pretty cool little tremolo effect for our oscillator. As janky as this may seem, a good 70 percent of guitar tremolo pedals operate on this flashing light–into–photoresistor principle.

After a little while, you might get bored of the staccato square-wave modulation source. What if we could make that filter's cutoff glide a bit more freely between values? Let's add a large capacitor between the LED and ground. Now, when the capacitor discharges, we'll see the light in the LED linger for just a little bit (Figure D).

Adding this capacitor changes the modulator's shape significantly (Figure E).

D

E

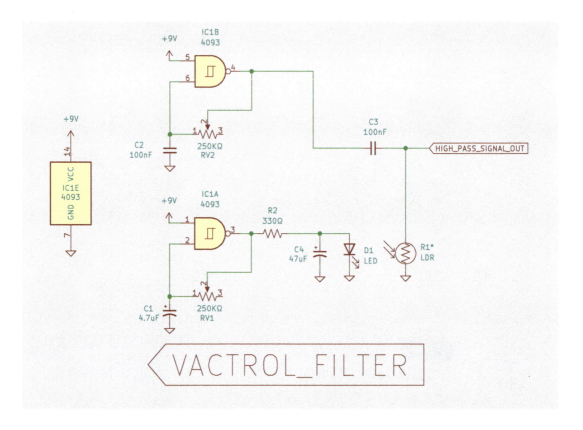

Above is the schematic of this whole setup.

When you've made a cool-sounding modulation source, it's best to seal up your creation with a little bit of electrical tape or heat-shrink tubing. Covering your photoresistor and LED in something opaque will help prevent unwanted light from modulating your filter.

This device is called a *vactrol* or an *opto-isolator*. Despite the fact that it's relatively basic concept-wise, this is actually a super-important component. Vactrols were developed in the 1960s by an engineer named Don Buchla on Dogbotic's home turf of Berkeley, California.

You can make one of my favorite little drone machines using two CD4093s. One IC is set up to have four audio oscillators, with each output heading into a photoresistor-controlled high- or low-pass filter. The second IC is set up with four flashing LEDs operating rather slowly. By positioning the LEDs near the photoresistors (and setting the LEDs to pulse at different rates), we can create a circuit that slowly evolves a droning musical texture. It's quite lovely—try it out.

PROJECT: PULSE-WIDTH MODULATION

MATERIALS:

- 2 x 10 kΩ resistors
- 2 x 100 kΩ potentiometers
- 2 x 1N4148 diodes
- 47 nF capacitor
- 4093 IC

Another interesting musical attribute to play around with is *pulse width*, or "how long your squares in your square wave are." We briefly touched on this with the Undertoner back in our "Chaining Oscillators" chapter. When you change the ratio of high to low voltage within a square wave, you'll hear the timbre of the sound change in a distinct and beautiful-sounding way. Take note that changing the pulse width doesn't change the frequency of the wave: it's still switching from on to off the same number of times a second. All that has changed is the duration of the "on" phase (Figure **A**).

Thanks to the one-way electron flow a diode generates, we can use this configuration to choke the signal before the full cycle finishes. Moving the potentiometer changes the pulse width, turning a rich and bassy square wave into something metallic and tinny.

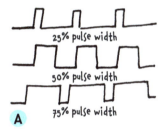

A

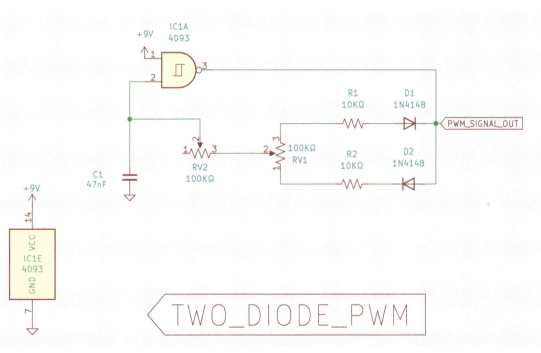

MATERIALS:

- 2 x 330 Ω resistors
- 2 x 10 kΩ resistors
- 2 x 250 kΩ potentiometers
- 2 x 1N4148 diodes
- 47 nF capacitor
- 4.7 µF capacitor
- 47 µF capacitor
- 2 x LEDs
- 2 x LDRs
- 2 x 1N4148 diodes
- 4093 IC

PROJECT: PULSE-WIDTH MODULATION WITH VACTROLS

Twisting a potentiometer is fun and all, but we can automate the process using the vactrol we just built. By using *two* photoresistors in place of a potentiometer and two LEDs as a modulation source, we can move the pulse width around at whatever rate we please.

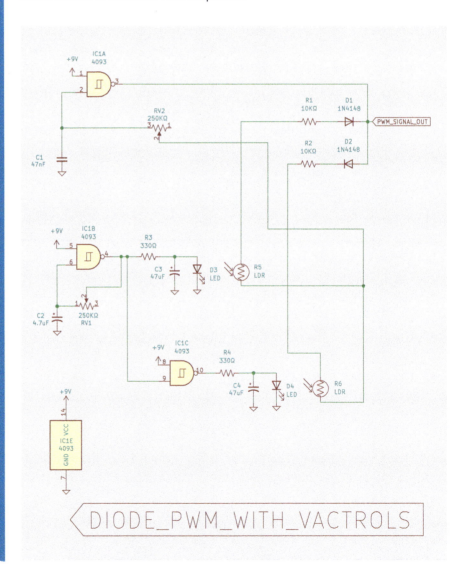

PROJECT: A BUTTON-CONTROLLED VCA

MATERIALS:
- 1 kΩ resistor
- 2 x 10 kΩ resistors
- 50 kΩ potentiometer
- 100 kΩ potentiometer
- BC548 transistor
- 2 x 1N4148 diodes
- Tact switch
- 100 nF capacitor
- Electrolytic capacitor of 1 µF to 470 µF in value
- 4093 IC

The *envelope* of a sound, generally, is how its volume changes over time. Imagine the sound of a violinist plucking a string versus bowing it. While both sounds are of the same frequency and share a harmonic spectrum, their envelopes are different. A plucked string is a short, fast sound—it suddenly happens, and before you know it, it's over. A bowed string, however, might slowly emerge from silence over several seconds and disappear just as slowly. Musicians use the terms *attack* and *decay* to refer to how quickly a sound comes in and goes away, respectively. By changing attack and decay times of the volume, we can make our synthesizer resemble plucks, bowed strings, percussive hits, and even brass.

This voltage-controlled amplifier (VCA) circuit is a wonderful opportunity to introduce you to yet another member of the electronic-component family: the ever-popular transistor. In this circuit, our transistor is going to act as a floodgate that's holding back charge coming in from our battery. When the transistor's second leg senses our input voltage, it will open and close the floodgate proportionally. This will result in the third leg giving off a battery voltage-level signal in the same shape as the envelope.

The circuit might look new, but it really is quite simple.

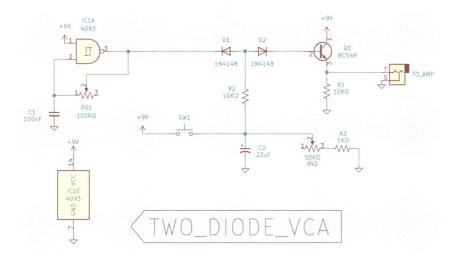

The heart of our VCA is two diodes that face away from each other (D1 and D2). If you recall, diodes allow current to flow only in the direction they're pointing—this means no current will be able to pass from our oscillator to our amplifier.

But why, you ask? Well, between these two diodes sits a capacitor that drains to ground (C2). The amount of charge in this capacitor determines the voltage between the diodes. When we press our button (SW1), we fill the capacitor with charge, therefore increasing the voltage at the junction. As our square wave oscillator wiggles around, it sucks voltage out of our capacitor. While no charge can move from the oscillator to between the diodes, the backwards diode *does* allow for charge from the capacitor to flow back into the oscillator.

This results in something interesting—the signal at our diode junction (and, thus, the signal flowing into the transistor) is the *mirror image* of the signal coming from our oscillator! Our amplifier will play the mirror of the AC signal that follows the contour of the capacitor decaying. Capacitors decay exponentially, which works well as an amplitude envelope. When the capacitor is fully drained (i.e., at 0 V), it stores no current to be mirrored, thus the output is silent.

PROJECT: AN LFO-PINGABLE VCA

Naturally, after being able to control the contour of your sound with a button, surely you'll say, "Why not an LFO? Couldn't we have an oscillator press the button over and over?" The answer is yes, but it requires an ounce of fiddling. The act of making one circuit's signal noticeable by another, we might call *conditioning.* Here, we will condition our square wave LFO to talk to our VCA circuit.

MATERIALS:

- 1 kΩ resistor
- 2 x 10 kΩ resistors
- 50 kΩ potentiometer
- 2 x 100 kΩ potentiometers
- BC548 transistor
- 4 x 1N4148 diodes
- Tact switch
- 2 x 100 nF capacitors
- 10 µF capacitor
- Capacitor of 1 µF to 470 µF in value
- 4093 IC

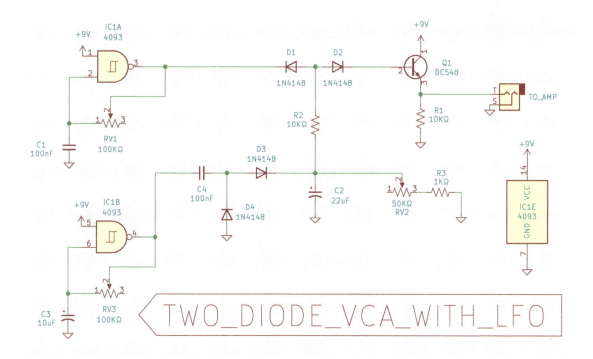

TWO_DIODE_VCA_WITH_LFO

- 1x piezo
- 2 x 1 kΩ resistors
- 2 x 10 kΩ resistors
- 1 MΩ resistor
- 250 kΩ potentiometer
- 50 kΩ potentiometer
- 100 nF capacitor
- 10 µF capacitor
- 100 µF capacitor
- 4 x 1N4148 diodes
- L7805
- BC548 transistor
- TL074 IC

PROJECT: THE PIEZO DRUM TRIGGER

One of our personal musical heroes at Dogbotic is the not-so-famous Jim Mothersbaugh, the original drummer for the left-of-center rock band Devo. Jim is better known as an engineer, and it was his stint with Devo that pushed him along that path. Encouraged by his brother Mark to create percussion that would sound like "bombs dropping," Jim rigged up what could potentially be the first electronic drum kit, and proceeded to baffle and piss off bands all over central Ohio.

How Jim's invention worked has been lost to history, but we were inspired to come up with a solution that at least provided the same effect. We're going to rig up a drum (or something we can thwack) with a piezo, which will generate a tiny voltage whenever we hit our drum. That tiny voltage will be sent through an amplifier to exaggerate the signal's strength, and then into an envelope generator.

This project works excellently with a square wave. (Hitting your piezo will result in the VCA applying a volume contour to your oscillator.) However, it works just as well with the circuits that we'll be presenting to you in the "Electronic Percussion" chapter (see page 230). That way, hitting the piezo on your drum will result in something that sounds kinda drumlike.

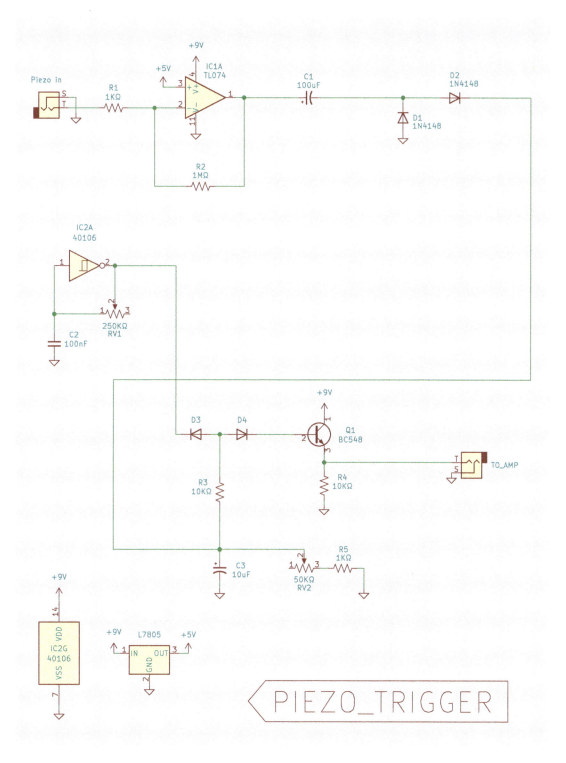

PIEZO_TRIGGER

MATERIALS:
- 2 x 250 kΩ potentiometers
- 4 x 1N4148 diodes
- 2 x 100 nF capacitors
- 4093 IC

PROJECT: RING MODULATION

A few chapters ago, when we first learned about filters, we discussed two ways you can *synthesize* a new sound out of two or more waves. We can add waves together (like a Telharmonium does) or subtract one from another (as with any filter-based instrument). You can, amusingly, also multiply two waves together and listen to the product. This process is called *ring modulation* because—well, take a look at the circuit.

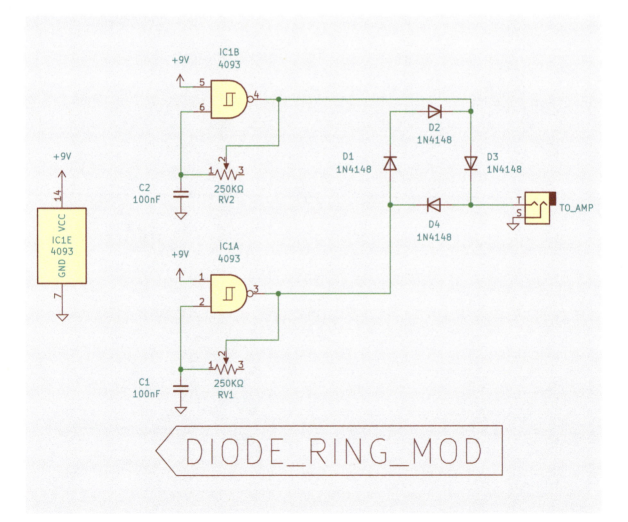

PROJECT: STEREO TREMOLO

MATERIALS:
- 2 x 100 kΩ potentiometers
- 4 x 1N4148 diodes
- 100 nF capacitor
- 10 µF capacitor
- 4093 IC
- 4066 IC

Another classic studio effect that's not too hard to emulate is ping-pong stereo panning—a trick in which a sound bounces back and forth between two speakers at whatever rate you desire. As we talked about when we built a stereo panner back in our amplifiers, *stereo* simply means "two independent signals coming from two speakers."

We've already seen how easy it is to make a stereo speaker system—we just need two amplifiers. In order to make the ping-pong effect, however, we must create a signal that will turn on one of your speakers while turning off another. We can do this thanks to the handy-dandy 4066 chip.

The CD4066 is a switch—it's surprisingly unfancy. Let's take a look inside (Figure **A**).

We see that the CD4066 has pins for voltage and ground and four momentary open junctions. Each of those gaps has another pin that, when provided with voltage, will close the gap. For example, pins 1 and 2 are normally isolated from one another, but if we connect pin 13 to positive voltage, pins 1 and 2 will connect.

Let's say pin 1 of our 4066 is our audio input and pin 2 of our 4066 is what's going to our left speaker amplifier. If we were to provide a square wave LFO to pin 13, we would hear our oscillator beep on and off with the timing of our square wave. This is half of the stereo tremolo effect—the tricky part is to produce another speaker that lets the audio through on the *inverse* of what our square wave is doing. Fortunately, we literally already have something that can be used as an inverter on our breadboard—our CD4093 NAND gate! If we can flip the square wave upside down, we can use *that* square wave to trigger another junction on the CD4066.

To use a NAND gate as an inverter, we must tie one of the input pins to positive voltage. Now, whatever the other input pin receives will be inverted and sent to the output. Any positive signal sent into the input will result in 0 V from the output, and any lack of current at the input will prompt the output pin to crank up the voltage.

Let's run our square wave through an inverter on the 4093. Hooray! Now we have two square waves, one entirely out of phase with the other: when one square wave turns on, the other turns off. See the schematic on the next page.

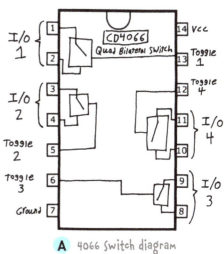

A 4066 switch diagram

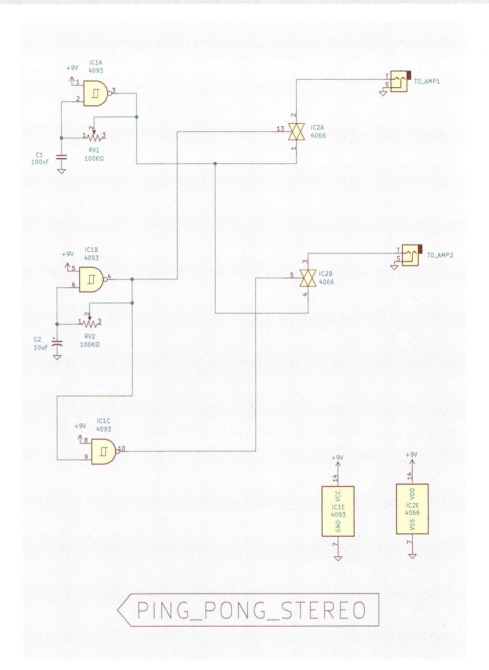

Play audio out of your two speakers while playing with the control oscillator's potentiometer. Changing the speed of the oscillator should result in a faster ping-pong effect. ♪

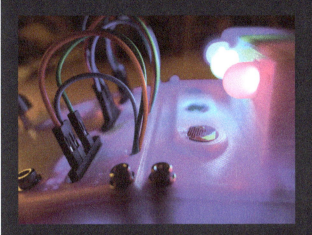

Musical Innovation: The Raindrop

Artist: Ramona Sharples

About: This synthesizer lets you compose melodies with patterns of flashing lights! Ramona was inspired by the way a guitarist can bend the strings of their guitar to warp the notes they're playing, and she wanted to capture that expressive, chaotic energy in a digital synthesizer.

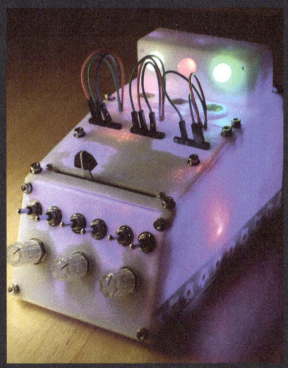

The Raindrop has three oscillators, and the frequency of each one is controlled by a combination of a knob to set the base pitch and a photoresistor that provides extra modulation. LEDs shine on the photoresistor and control the pitch! There's also a set of switches that let you turn the notes and LEDs on and off.

The leftover fourth oscillator on the quad-NAND chip acts as a clock signal for a set of three 4040s. The divisions of that clock frequency from the counters are available in little headers in the middle of the synth, and the player can use provided patch cables to connect the red, green, or blue pins on each of the LEDs to any of those clock signals. This creates a pattern of colors in the LED that in turn creates a little melody! The player can further change the melody by shining a flashlight or covering up the LEDs to change the amount of light hitting the photoresistors.

11

SEQUENCERS

A *sequencer* is a device that reads back a series of steps in—you got it: a sequence. Pat yourself on the back.

A sequencer is perhaps the most straightforward synthesizer part we've mentioned so far. A lot of people think about sequencers as little computers that can be programmed to play back a short little melody: picture eight potentiometers, each controlling a note in an eight-note musical pattern. While the sequencer can "read through" those eight potentiometers one at a time, essentially giving you the power to preprogram a musical idea, that information *doesn't have to be pitch*. You could also use a sequencer to program eight filter cutoffs and rhythmically jump between them.

We use sequencers to tell electronics *when* to change a parameter. Sequencers pop up everywhere in our built world, not just in electronic music. Take the traffic lights at your local intersection: They have a preprogrammed sequence—sort of like an electronic script—that they follow religiously. One light turns red, we experience a short pause, and then another light turns green. Having a reliable sequencer is vital for any intersection—otherwise, two cars might get conflicting signals at the same time.

A musical sequencer is surprisingly similar to the device that manages traffic at your nearest crossroads: by adjusting a series of potentiometers, we can program our sequencer to play back a series of specific voltages over and over again. A classic use of this is to play back an eight-step musical phrase at blinding speed. Think of any classic disco song that features a continuous and rapid series of notes—that part was very likely performed not by a human with a keyboard but by a human who programmed a sequencer. With a sequencer, you're not just a musician with an instrument—you're a conductor with an orchestra.

You might have heard your digital audio workstation referred to as a sequencer. It is! It's just a sequencer with a *really, really high number of steps*—so high, in fact, that you don't even think about it as a step sequencer. But make no mistake—every MIDI note played is indeed read back in a sequence.

A sequencer can be used for much more than just playing back pitches, though. Because a sequencer is simply calling back a series of voltages, we can use those voltages for any variety of nefarious purposes. For example, we could use the sequencer to change our note's decay length or our filter's cutoff frequency. Much like an LFO or an envelope generator, a sequencer doesn't produce *sound*—it produces *control voltage* that will be used to change a parameter.

Most musical sequencers have four, eight, sixteen, or thirty-two steps because most humans have two feet. If we lived on a planet where people had five feet, we would probably listen to music where sequencers are set to five, ten, twenty, or forty steps. Fortunately, because you're building your own sequencer, you can pick however many steps you want! With a greater number of steps, however, comes greater complexity. Start off building a small number of steps, and only once you have something working should you consider scaling the project up.

THE 4051

To build our sequencer, we'll need to employ our most complex IC yet—the 4051 multiplexer. Don't mind the supervillain-esque name—it's really not too bad. A multiplexer is, in theory, a simple device—it's a selector that can route one wire (the *common wire*) to eight other wires (the *peripherals*). If we send power into our multiplexer's common wire and then tell the multiplexer which output to send it to, we can selectively light up one of eight lightbulbs attached to the multiplexer's peripherals. We can't light up more than one lightbulb at a time, but we can switch which bulb receives the current as quickly as we like.

Let's discuss a more amusing musical usage of this. If we play our favorite Björk record into our multiplexer's common wire and attach each of our eight outputs to a speaker, we can make Björk's voice jump around the room from speaker to speaker at blazingly fast speeds. All we'll have to do is tell the multiplexer to move between the outputs really quickly.

Conversely, we can do the exact opposite! Instead of using the common wire as the input and the peripherals as outputs, we can swap them. Now, instead of one Björk record, we get eight CD players and send eight of Björk's greatest Icelandic hits into each of the peripherals. If we plug a speaker into the common wire and tell the multiplexer to quickly move between the inputs, we'll hear a collagelike channelsurfing extravaganza as we rapidly jump from Björk song to Björk song.

A multiplexer is just a connector—nothing more. Because of this, we can use it as a way to send one signal to up to eight places or send up to eight signals to one place. Imagine the convenience! For a pleasant railroad analogy, which I'm partial to, I recommend thinking of a railway interchange. These things resemble a record player turntable, with a set of tracks across its diameter. A locomotive can be pushed onto the turntable, and the turntable can

Adobe Stock-Semi

A rotary turntable for trains is similar to our multiplexer circuit. It can connect a train on one track to many potential new tracks. Our IC connects a single channel (Pin 3) to many other channels (Pins 1, 2, 4, 5, 12, 13, 14, and 15)

rotate so the locomotive can be put on another set of tracks. This is a mechanical multiplexer: it allows one input to go to one of many outputs, or one output to go to many inputs.

Check out the drawing of the 4051 on the opposite page. The power pins are exactly where you would expect—pin 16 is for your 9 V connection and pin 8 is for your ground. Pin 3 is your common input. Pins 1, 2, 4, 5, 12, 13, and 14 are our peripherals. Note that the peripherals are not consecutive, because whoever designed this IC had a personal vengeance against folks learning electronics.

You'll notice I haven't yet told you exactly *how* you tell the multiplexer which pin to connect to the common pin. Pins 9, 10, and 11 are called our "address" pins. In order to know where you're going, you need to know the address! We can use these three pins, and a little bit of math, to communicate with this IC. But to do this, we're going to have to learn how to speak Machine.

BOOLEAN LOGIC

Ten is an important number for us humans, based exclusively on the number of fingers most people have. In the intervening twenty million years since the development of the primate hand, humans have taken their obsession with the number ten and run with it. Now, for some reason, celebrating your thirtieth birthday is a much bigger deal than celebrating your twenty-ninth, and doing forty crunches somehow just feels more satisfying than doing forty-one.

Among the most annoying side effects of the human fixation on ten is that it led to a funny number system. We have ten symbols that can be used to express absolute quantities of anything. You've probably seen them—they look like this:

0 1 2 3 4 5 6 7 8 9

Because we have ten symbols, we call our system base ten. Once you get to the number 9 and you run out of symbols, all you have to do is add an additional space for a second digit: After 9 comes 10. After 99 comes 100. Because humans have ten fingers, we call our number system *decimal logic*.

Computers don't have fingers. In fact, computers don't really have any *things* to count on. Inside your computer are millions of tiny wires, and at any given point in time, each wire can either have a tiny bit of current flowing through it or absolutely nothing at all. Just like the 4017 from our harmonization chapter, a computer uses current like we use our fingers—when nothing is flowing, it represents a 0 (as if we had all our fingers collapsed into a fist). When current flows, it represents a 1. Because a computer can count only two numbers, we call this system *binary logic*.

Any individual number in binary, using a 0 or a 1, can be referred to as a "binary digit," or "*bit*." When you have eight bits together in a row, we call that a *byte*.[1] A kilobyte has 1,024 bytes, 1,024 kilobytes make a megabyte, 1,024 megabytes make a gigabyte, and so forth. If you sat down with a big sheet of paper and wrote out 8,589,934,592 bits—that's eight and a half billion individual 0s or 1s, you would have written out a gigabyte of information by hand.

You've certainly, at some point, heard that computers use 0s and 1s to communicate. You probably smiled and nodded and then hoped nobody would ask you questions about it *because it makes no sense*. I always thought this was a confusing phrase, because your computer doesn't literally have little numbers running around in it. Your computer is nothing more than a very complex switch with a lot of wires going into it. When the correct combination of input wires

1. When you have four bits in a row, we call that a *nibble*. The nibble isn't really ever used in computing, but it is a cute joke, so we chose to leave this tidbit in.

are turned on, the output wires respond in some preprogrammed logical way. To communicate a signal, a wire in your computer flashes a signal in computer code. To make it easy to talk about it, we label the "flashes" of current the number 1 and the "pauses" the number 0.

To communicate a number larger than 1, we're going to have to learn how to count in a base 2 system. But don't worry—base 2 is just like base 10 if you're missing eight fingers.

We start counting with a 0, then a 1, and then, oops—we're out of digits. So, just like when we reach 9 in decimal logic, we add a new place.

0
1
10
11
100
101
110
111
1000
1001
1010
1011
1100
1101

And so on and so forth. You'll notice there's a tradeoff—while a binary system has fewer symbols to worry about, you'll have to use proportionally more digits to communicate a number in binary than in decimal. For instance, a single digit in decimal (9), takes up four times as much space in binary (1001).

BINARY AND THE 4051

The 4051 is used to thinking in binary. We can connect each of the address pins to either a voltage source (to communicate a 1) or to ground (to communicate a 0). If we tell the 4051 the binary number that corresponds to the number of the peripheral we want to connect to, the IC will do so without hesitation.

For example, let's say we want to send a signal from our common input (pin 3) to the first peripheral (marked as number 0). A three-digit representation of decimal 0 in binary is, unsurprisingly, 000. We then hook up each of the address

Bases Loaded

A question that gets asked often in our workshops is, "Why base 2?" Wouldn't it be possible—and much more efficient—to build a computer that works in base 3? Instead of just "on" and "off," we could also have an "in the middle" voltage amount. I love this question because *it's totally possible*—and in the 1950s, the Soviets decided to do just that. They quickly ran into two key problems.

REASON 1: CONFUSION

It's fairly easy for a circuit to tell whether current is there or isn't. Introducing a "middle" voltage makes it considerably easier for the circuit to get confused. What if that middle voltage slightly overshoots and is read as full voltage? The voltage difference on this scale is truly tiny—millivolts of potential energy—which leaves very little room for error.

REASON 2: CONVENIENCE

We already had a lot of research on doing math in base 2, and virtually none on base 3. And so it stuck.

As of this writing, in 2024, there is ample research in using DNA as a means of saving digital information. DNA is incredibly tiny, has built-in error-correction systems, and has four nitrogenous bases (A, T, C, and G) and thus can operate in base 4. We're a ways off from a world of DNA-based memory cards, but keep an ear to the ground. Some day, your phone might be encoded with the very language of life. Spooky.

The Сетунь (Setun), one of the valiant Soviet attempts at making a base 3 electronic computer

pins to ground to represent the 000, and voilà—our signal goes into pin 3 and out of the pin labeled Q0 (pin 3).

Neat! Let's switch the signal to the output peripheral marked Q3. A three-digit representation of decimal 3 in binary is 011. We hook up our address pin A to ground and address pins B and C to +9 V. Immediately, current flows from the common pin to peripheral Q3.

"Now what?" I hear you say. We expected to make a *sequencer*, not some sort of silly switch with the ease of use of a phone switchboard. How can we make this *fun*?

I'm so glad you asked. The joy of the 4051 is that it's an *electronic* switch, which means we don't have to physically connect it elsewhere. We can make a signal that counts up to eight over and over again.

Remember our old friend the 4040? We used it in the previous chapter to make octaves and to power our vactrols. Now, we're going to use it *in the way it was intended to be used*. If you recall, our 4040 takes an input signal from a square-wave oscillator and bifurcates it several times over into signals that might look like this:

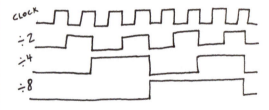

What I'm about to show you is perhaps one of the most single beautiful, serendipitous, and utterly shocking things in this entire book. Way back when we were discussing our first oscillator project, we talked about how a voltage signal can represent a sound by pushing and

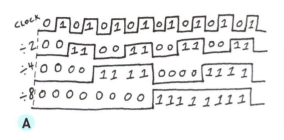

A

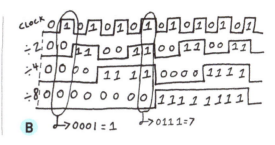

B

pulling our speaker cone so that it changes air pressure in the same way a "real" sound would. A few chapters later, we talked about how voltage signals can also represent a *control* signal—we can use a square wave, for example, to modulate volume or a filter's cutoff. Now, we've learned a third thing that a changing voltage can represent: digital data.

Instead of thinking about the squares above as analog waveforms, we can write in their Boolean expressions—if the signal at any point represents a 1 or a 0. Stare at Figure **A** until you notice something interesting.

Let's look at these waves vertically, and read them as columns from top to bottom. We first see 000, then 001, then 010, then 011. We are in fact, counting in binary (Figure **B**).

At the heart of every digital computer is an oscillator going into a binary divider—just like your square wave and your CD4040. The binary divider converts a pulse into numbers a computer can count. You are about to make, without exaggeration, a digital computer on a breadboard.

The first step to building a sequencer is to set up three ICs in a row—a 4093

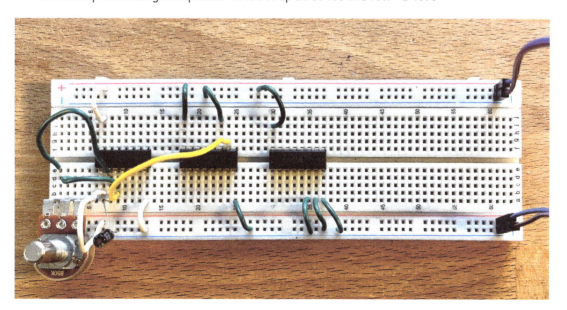

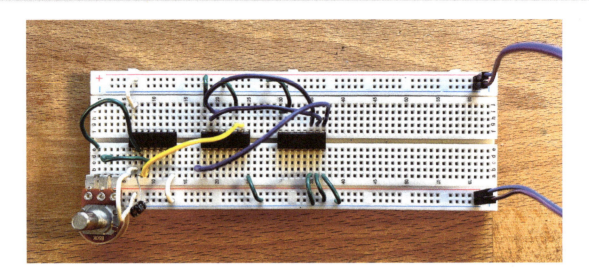

(our oscillator), a 4040 (our divider), and the star of the show: a 4051 (our multiplexer). Connect 4051 pin 16 to supply voltage and pins 6, 7, and 8 to ground.

If we attach three consecutive pins from the 4040 to our 4051's address pins, we will, in essence, be telling the 4051 to count from one to eight over and over again. Every time we change our number, we change the connection going through the 4051. Now we don't have to manually rewire the address pins to change the output—we can do so electronically!

PROJECT: THE DISCO BOOLE (PART I)

MATERIALS:
- 100 kΩ potentiometer
- 100 nF capacitor
- 4093 IC
- 4040 IC
- 4051 IC

The easiest sequencer to start out with is one I affectionately call the Disco Boole. It's capable of playing any iconic synth bass line you'd like, but the only catch is that it can play bass lines only if they're from 1982.

The Disco Boole, really, is just an octave sequencer. Set your oscillator to a frequency you like, and the Disco Boole will play back that note in a bunch of octaves, one at a time. It doesn't sound terribly interesting when you put it that way, but I promise it's surprisingly groovy:

As you might have predicted, our breadboard image is quite noodle-y. Let's take a slightly more oblique view of the breadboard so we can better see what's going on. Note the oscillator going into the 4040 and the three

"address" wires (in blue), which talk to the 4051 in binary. All that remains are the yellow wires, which connect from different suboctaves of the 4040 to the switch in the 4051.

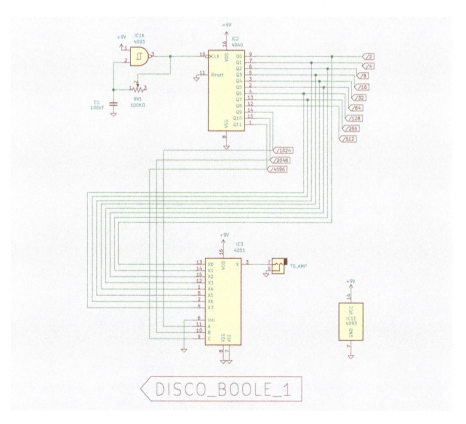

PROJECT: THE DISCO BOOLE (PART 2)

MATERIALS:
- 2 x 100 kΩ potentiometers
- 100 nF capacitor
- 4.7 µF capacitor
- 2 x 4040 ICs
- 4093 IC
- 4051 IC

The Disco Boole uses only one oscillator, so the pitch and tempo of this sequencer are inextricably linked. If you want to speed up the sequence, you must also increase the pitch of the oscillator. How could we give the Boole separate pitch and rate control?

The answer, of course, is pretty easy: Create a second oscillator, add a second 4040 binary divider, and use *that* as your address pin input. Now, one 4040 will count in binary at whatever speed you'd like and the other 4040 will provide the suboctaves.

MATERIALS:

- 8 x 100 kΩ potentiometers
- 250 kΩ potentiometer
- 8 x 1 kΩ resistors
- 100 nF capacitor
- 10 µF capacitor
- 4040 IC
- 4093 IC
- 4051 IC

PROJECT: THE STANDARD EIGHT-STEP SEQUENCER

The Disco Boole leaves a lot to be desired, of course. How could we build a sequencer that, you know, has the ability to tune each note?

With just a few more potentiometers (and a lot more wires), we can make this dream of ours a reality.

Our 4093 oscillator can be tuned by changing the resistance between the NAND's input and output pins. If we had really fast hands, we could play a sequence by pulling out potentiometers and reinserting new ones at lightning speed. But what if we could have the 4051 do that? What if each pathway through the 4051 went through its own potentiometer, thus instantly tuning the oscillator to whatever the dialed-in resistance is?

This method is pretty nifty. We need only two oscillators to get this working. One oscillator drives the 4040, which counts the 4051—the other is the one that's getting retuned. In fact, the only breadboard eyesore you'll encounter is stringing up eight potentiometers.

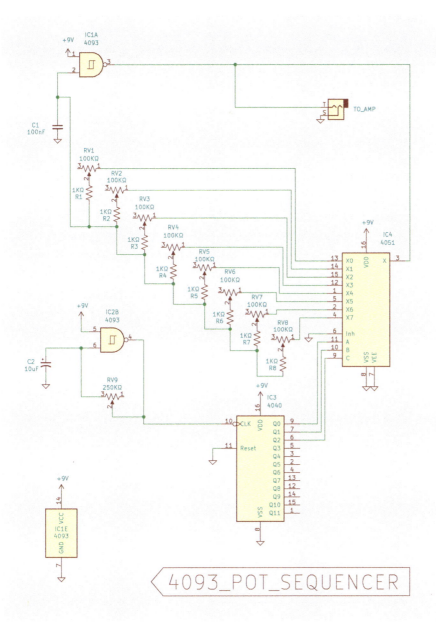

4093_POT_SEQUENCER

What if you don't want eight notes in a row? What if you had rests? An easy way to be able to turn notes on and off is to attach an SPST switch between the 4051 and its respective potentiometer output. By cutting off the current before it hits the potentiometer, the oscillator won't feed back into itself and, thus, won't make any sound.

MATERIALS:

- 8 x 100 kΩ potentiometers
- 250 kΩ potentiometer
- 8 x 1 kΩ resistors
- 100 nF capacitor
- 10 µF capacitor
- 4040 IC
- 4093 IC
- 4051 IC
- 3 x SPDT switches

PROJECT: A PATTERN-CHANGING SEQUENCER

A lot of fun can be had with this sequencer by messing with the 4051 address-pin inputs. Plugging new 4040 subdivisions into the 4051 will result in all sorts of fun secondary patterns. A row of tact switches can make this more playable—by tapping a button, you change the signal going to an address pin and, thus, the sequence. You could even rig three oscillators, going at their own speeds, as the three binary counters needed for the three address pins.

The schematic on the next page shows one of these variations—three SPDT switches allow you to toggle between different binary outputs. When any switch is flipped, the pattern of notes is played by the sequencer (though not the frequencies of the notes). This is marvelously playable, as a single switch flip can result in a handy variation of your rhythm. Each combination of switches will likewise produce a new pattern.

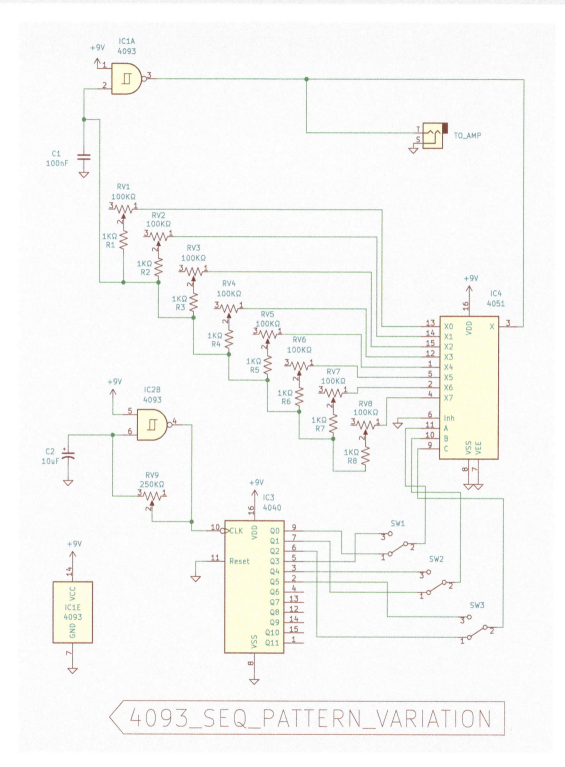

4093_SEQ_PATTERN_VARIATION

MATERIALS:
- 100 kΩ potentiometer
- 250 kΩ potentiometer
- 100 nF capacitor
- 10 µF capacitor
- 4040 IC
- 4093 IC
- 4051 IC
- 4017 IC

PROJECT: A FIRST-ORDER RESET HARMONIZATION SEQUENCER

Perhaps our favorite schematic in this entire book is this rather understated one. This is a sequencer that uses both a 4051 and a 4017 decade counter from our harmonization chapter (see page 174). Instead of sequencing individual potentiometers, this circuit sequences *where the 4017's reset pin is plugged in*. This will result in a different musical interval sounding every time the 4051 moves pins, in harmony with the oscillator's pitch.

We call this the reset sequencer because the magic comes from the fact that one of the pins on our 4040 will be used to "reset" the count on our 4017. By resetting a decade counter at slightly different intervals, we can make the 4017 sing a series of different pitches. Because all the reset intervals are integer divisions of 2, the resulting sound is a self-generating melody that doesn't play any notes out of tune.

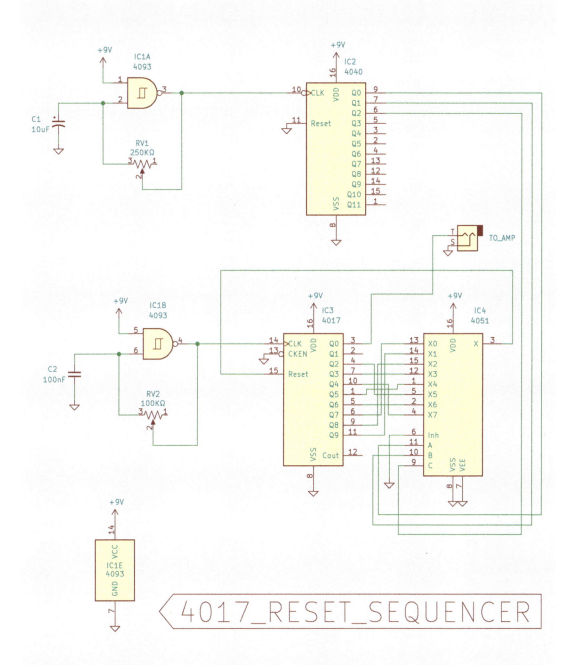

4017_RESET_SEQUENCER

MATERIALS:

- 100 kΩ potentiometer
- 250 kΩ potentiometer
- 2.2 kΩ resistor
- 100 nF capacitor
- 10 µF capacitor

PROJECT: A SECOND-ORDER RESET HARMONIZATION SEQUENCER

I know what you're thinking: How could we possibly make this more complex? Well, what if you want this automatic sequencer to stop at a number shy of eight? The 4040 has a reset pin—just like the 4017—and can be set to an arbitrary number below eight with a clever diode trick. Just like relays, vacuum tubes, or transistors, diodes can be used to model logical operations. In this case, we're going to use two diodes to model the operation AND.

A diode AND gate looks like the image to the right.

This circuit is pretty simple—current flows through a resistor, into an LED, and *out* two diodes. On the other side of the diodes sit two inputs, labeled I1 and I2. If we were to hook these inputs up to VCC, the charge on both sides of each diode more or less equalizes. Because there is now very little voltage across each diode, the current instead flows full force through the LED to ground. If only one of the diodes is connected to VCC, the other diode provides an escape for the extra current.

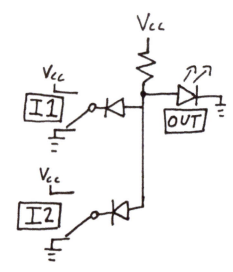

Let's say we want our sequencer to reset at the number 7. Seven, represented in binary, is 111. This number is represented with electric charge on our 4040 pins 9, 7, and 6. We can use diode logic to reset the 4040 when pin 9 *and* pin 7 *and* pin 6 are all expelling charge. Simply use a diode to connect each of those pins to the reset pin (pin 11).

We can reset this to a number other than 7, too. Just write your number in binary. For every 1 you see, connect that place's respective pin to the reset pin through a diode. When that combination of 1s pops up, your diode logic will instantly start counting over from 0. 🎵

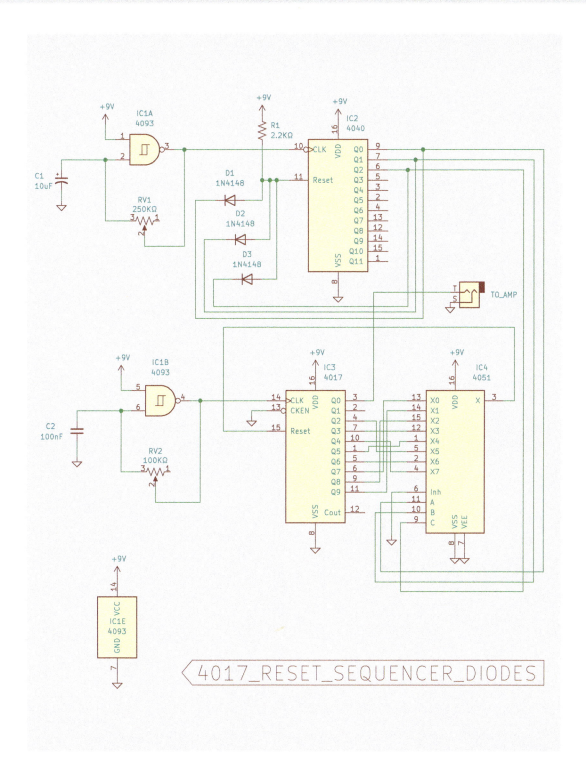

Musical Innovation: Neon Mushroom Garden

Artist: Jasmine Bailey

About: From the artist herself: "I love mushrooms! I love LEDs and making them blink! I love making things from start to finish! Enter the Neon Mushroom Garden—a small series of dioramas I made featuring the 4093 oscillator I learned about in Dogbotic's DIY synthesizer workshop. I decided to use all four oscillators so that I could make a nice twinkle effect with a little bit of variation. In the black-and-white diorama, I also included LEDs inside the sculpted mushroom stems! In this way, it operates a bit like a night-light, where one switch turns the mushrooms on and another switch adds the twinkle effect. In my next diorama, I hope to include the cricket synthesizer and spring reverb I learned about in the same workshop to incorporate more of the senses!"

et's set the scene: You're a budding ceramicist living during the Han dynasty. Despite your vessels being of the utmost quality, your business is under risk from a hot new trend taking the youngsters by storm: metalwork. We're talking vessels forged out of bronze, with all sorts of fancy features like side rings for hanging, or bumpy rivets for fastening. While clay pots don't *need* any of those things (you don't rivet clay—that's not how that works), you realize it's a heck of a lot cheaper to fashion pottery to *look like metal* than the other way around.

Turns out your plan was an excellent one. Everyone wants metal, but nobody can afford it. Now you're churning out clay pots that mimic newer, fancier pots. Clay, being literally dirt cheap, means higher profit margins. *Cha-ching*. Or whatever sound ancient cash registers made.

This isn't purely hypothetical—archaeologists really have found ceramics all over the world with faux-metal signatures, but this idea extends far beyond pottery. Here's a fancy design term for you: *skeuomorph*. It's pronounced like "skewer morf" with a Boston accent. A skeuomorph is a design feature that imitates an older version of itself, just for kicks.

Picture an electric light that's styled to look like a flickering candle: The light doesn't *have* to flicker or be in a candle-shaped body, but it's been engineered for the express purpose of doing so. It emulates candlelight in order to make you *think* of candles when you turn it on, without bothering with all those annoying features of real candles. Likewise, your analog-clock display on your computer is a skeuomorph, as is wood laminate on your grandma's tacky 1970s-vintage kitchen cabinets. The world is filled with things that harken back to older versions of themselves. It's an odd comfort, until you stop and think about how absolutely strange it all is.

Audio skeuomorphs exist, too. From 2007 to about 2017, every photo taken on a smartphone was accompanied by the sound of a mechanical shutter. Turn off the screen on the very same phone, and you'll hear the sound of a padlock closing. Does the sound of a lock make the phone any more secure? No, obviously not. But that phone's designer is betting that hearing that *click* makes you feel just a little more safe.

Sometimes skeuomorphs are actually quite vital, and not just for comfort. For instance, electric cars don't make nearly as much sound as gasoline-powered cars, which results in many unexpected accidents. Nowadays, most electric cars actively *play audio* to make up for their silence. These cars are

The Metropolitan Museum of Art

An ancient skeuomorph: pottery with faux-metal rivets.

A modern skeuomorph: a user interface's deletion tool resembling a trash can.

engineered with speakers on the outside! As it turns out, cars making sound is essential to the safety of people around them.

Skeuomorphs represent a strange human anxiety around technological change, and for this reason they're marvelous symbols to exploit for creative use! And there's no skeuomorph popular music loves more than the drum machine. Drum machines are magnificent musical instruments that combine the pattern-making capabilities of a sequencer with a number of sound sources that traditionally mimic acoustic drums.

A drum machine is an oddly powerful symbol:

- Analog drum machines don't play back recordings. An electronic drum sound is produced by a circuit that's being hit by a jolt of current, much like how a drum is hit by the jolt of a drum stick. Each drum sound is a circuit that literally vibrates electrons in a pattern that's shockingly similar to how a real drum vibrates air molecules. The circuits themselves are symbols of drums.

- The sounds of electronic percussion, while once startlingly novel, now represent a bygone era that never seems to go out of fashion. The majority of people making retro 1980s music were born during the administration of the second President Bush.

- Drum machines were once a reviled symbol for reasons that are largely misguided. For many people, electronic percussion seemed like a means to put drummers out of work. To a lot of people, it still represents the crushing force of capitalism. I say that's delicious, and you should use that to your creative advantage.[1]

A BRIEF HISTORY OF DRUM MACHINES

The Rhythmicon, invented in 1930, is frequently cited as the prototypical drum machine. Invented by a hero of experimental music—our old friend Leon Theremin—for composer Henry Cowell, the Rhythmicon wasn't much different from the harmonization circuits we built a few chapters back. Instead of an integrated circuit doing the divisions, however, Theremin used a metal disk with holes punched in it. The disk could spin at audio rate (making this an audio synthesizer), but it could also be played at subaudio rate (making it a magnificent clicking machine).

As we mentioned in the introduction, the idea of "firsts" gets truly muddy the

---

1. Read more about this social quirk in our final chapter, Thoughts On Automation.

more you have to define what something is. While a curiosity, the Rhythmicon remained an academic novelty project and was (largely) forgotten for about twenty years. However, the internet seems to be in agreement that this is the first electronic rhythm machine, and the internet has never been incorrect before.

In the postwar boom of the late 1940s, businesses tried to capitalize on electronic rhythm machines by insisting they were the special sauce you needed to kick your church group up a notch. For example, Harry Chamberlin produced the Rhythmate, a machine the size of a small refrigerator that could play back ten—yes, ten—dorky-sounding rhythms on independent tape loops. Although you could change the tempo, you couldn't edit the pattern (the Lord did not intend it).

Widespread adoption of the integrated circuit and the mass production of circuit boards came about during the 1950s. The Wurlitzer Sideman, from 1959, might be the simplest instrument that by today's standards I would confidently call a drum machine. A little like the Rhythmicon, the Sideman was controlled by a nifty clocklike mechanical sequencer: As the hand spun around the dial, it brushed over differently spaced metal contacts. Each time the hand touched a contact, it sent a pulse of current into a circuit that vibrated in a way that kinda-sorta sounded like a drum. To make the tempo faster, you just sped up the clock hand!

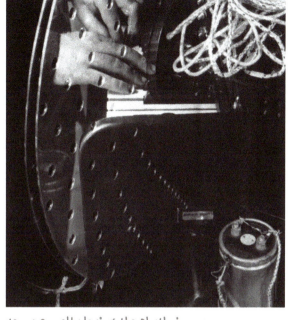

Henry Cowell playing the Rhythmicon, 1932

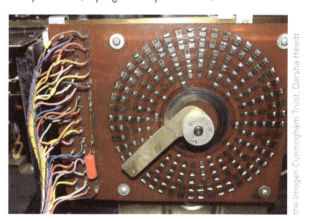

the Imogen Cunningham Trust, Darsha Hewitt

The internal mechanism for the Wurlitzer Sideman was an electromechanical sequencer. A clocklike arm would spin around a dial and close switches in different rhythmic patterns.

Unfortunately, you couldn't compose your *own* patterns—you were stuck with the deliciously cheesy factory presets (including the cha-cha, the shuffle, and something labeled "western"). You could, however, press individual buttons to trigger the drum circuits (i.e., you could make a drum roll by repeatedly pressing a button).

Eko's Computerhythm CR-78, from 1972, added programmability—the ability to make your own custom rhythms using an astounding sixteen drum voices. The LinnDrum, revealed to the world in 1982, went one step further, incorporating *recordings of real drums* as samples that could be triggered.

While the LinnDrum, arguably, created pop music's hunger for electronic percussion, it's where our story stops. The drum machines we'll be working with operate not with digital recordings but with simple circuits that resound slightly differently every time they're played. Interestingly, prior to the LinnDrum, *almost all drum voices operated with the same circuitry*. The circuits we're making in this chapter will certainly remind you of the sounds of numerous classic drum machines—that's because they're literally the same circuit.

THE 40106

The 40106's legal name is a hex Schmitt trigger inverter. It has six inverter units within it, and each inverter has an input and an output. It works almost like a 4093 without a control pin. Let's take a look-see at the image below.

Some of the circuits in this chapter involve copious numbers of oscillators, and the 40106 makes them one connection simpler. Now, you can fit six oscillators, not four, on a single chip—think of the savings! Though the schematics in this chapter all use the 40106, you can also use your 4093—or any other square wave source—to do the same thing.

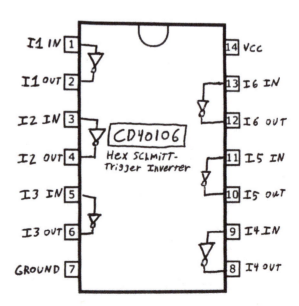

PROJECT: MAKING INHARMONIC SOUNDS

Back in our chapter on filters (see page 156), we learned about spectrograms and how they can be used to quickly glance at what a sound is composed of. Now, it's puzzle time, kiddos. Below are four spectrograms for four instruments. Which one is not like the others?

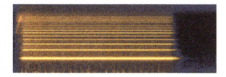

1. Accordion

2. Piano

3. Oboe

4. Crash cymbal

This riddle isn't very hard. It's number 4. A crash cymbal's spectrogram looks *nothing* like the others.

The first three instruments have pretty-looking harmonic spectra—look at those evenly spaced overtones! A cymbal, on the other hand, doesn't look that way at all. It's a rich spectrum, all right, but the partials are all over the place.

Certain instruments, like the accordion, or the cello or the clarinet, are meticulously shaped so they optimally excite even multiples of the fundamental. These instruments are great at making pitches you can hum along to. A cymbal, on the other hand, doesn't really produce pitch at all—it makes noise. A lot of percussion instruments are designed *opposite* from pitched instruments: instead of trying to evenly space their overtones, their superpower is the simple fact that their overtones are scattered.

To make an *inharmonic* (yep—not-harmonic) sound, we'll look to the CD4070B—a quad XOR gate. XOR is a logic gate, just like the NAND gates in our oscillators. *XOR* stands for "exclusive or" and has two inputs and one output. The output dumps current at full force when—and only when— one input receives current and the other input does not. If both inputs are

MATERIALS:

TAMBOURINES
- 4 x 250 kΩ potentiometers
- 4 x 100 nF capacitors
- 40106 IC
- 4070 IC
- Tact switch

CRASH CYMBALS
- 6 x 250 kΩ potentiometers
- 2 x 100 kΩ potentiometers
- 4 x 10 kΩ resistors
- 6 x 100 nF capacitors
- 2 x 10 µF capacitors
- 47 µF capacitor
- 2 x 40106 ICs
- 4070 IC
- 4 x 1N4148 diodes
- BC547 transistor
- Tact switch

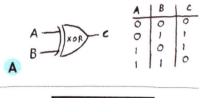

A

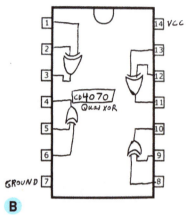

B

receiving current, or if neither receives current, the output stays dry.

We can model what an XOR does with the following truth table (Figure **A**). (*A* and *B* represent the inputs, and *C* represents the output.)

Taking a look at the chip, we find that it's called a *quad* XOR not because of its consistent leg routine but because the chip manufacturers were kind enough to give us four of these little marvels (Figure **B**).

You can use an XOR gate to make some fascinating sounds with unrelated square-wave inputs (e.g., two 40106 oscillators acting as *A* and *B*). When both square waves are in the same state, the XOR won't trigger. This will result in some very gnarly and inharmonic sounds—just you wait.

TAMBOURINES
INSTRUCTIONS:

It might be disingenuous to call this a tambourine, but you can take that up with the Roland Corporation. We think it sounds glorious (and it's a good beginner way to dip your toes into the world of drum machines).

Let's start with the basics—let's use a 40106 to make a square-wave oscillator. Shockingly, this is even easier than a 4093—all you need is a capacitor and a potentiometer (Figures **C** and **D**).

Next to it, let's include our 4070—following the standard power-pin arrangement.

We'll now connect two square waves to the inputs of the first XOR gate (Figure **E**).

Then we'll connect two more square waves to the inputs of the second XOR gate (Figure **F**).

Finally, we'll take both of those XOR outputs and attach them to a *third* XOR.

Let's take a listen to the output of the third XOR while playing with our square-wave frequencies. It should sound like two fax machines mating. When you hear a combination of frequencies that sounds particularly harsh, brittle, and awful, you've found it.

With the extra space on the breadboard, we can throw on our VCA circuit from the modulation chapter. Tune the decay potentiometer so the envelope matches the tambourine of your dreams. Just press the button, and voilà: your ticket to backing-band stardom. Check out the schematic on page 238 (Figure **G**).

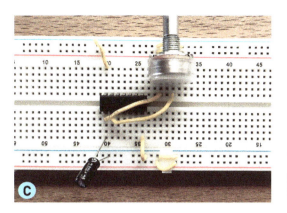

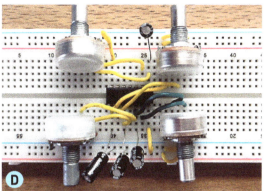

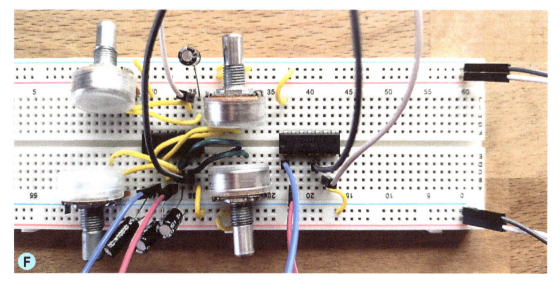

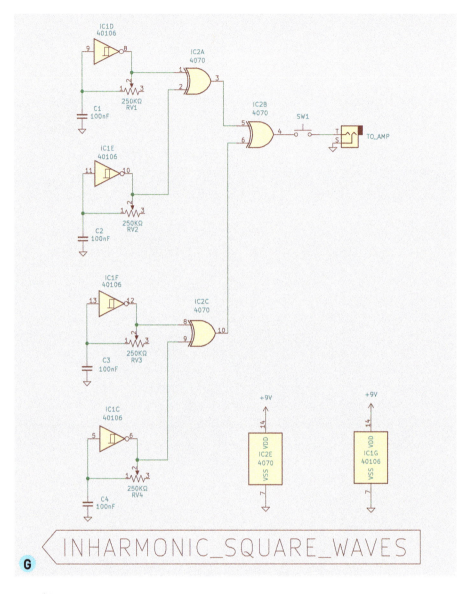

INHARMONIC_SQUARE_WAVES

G

CRASH CYMBALS
INSTRUCTIONS:

The difference between a tambourine and a crash cymbal is a second stage of XORing. If we use all six 40106 oscillators and four XOR gates, we can make something sound much more cymbally. Throw it into a VCA, and we've got ourselves a classic drum machine cymbal sound (Figure **H**)

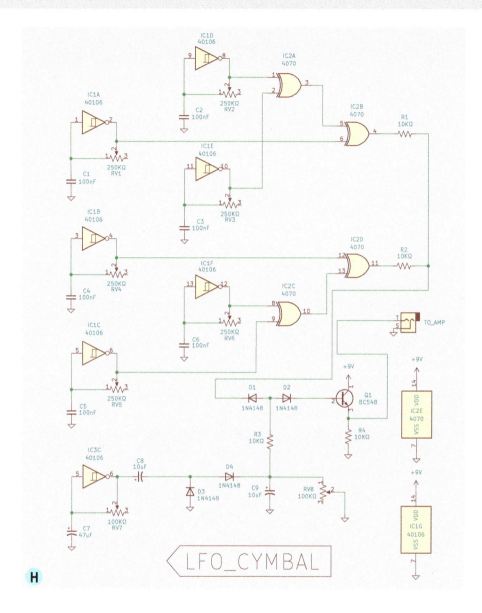

LFO_CYMBAL

H

Variation: Hi-Hats

Hi-hats are cymbals with pedals, which allows the drummer to modulate decay length. With a foot down, the closed hi-hat makes a staccato *"fizz."* With a foot up, an open hi-hat rings like a crash cymbal. To mimic a hi-hat, combine the crash cymbal schematic with a vactrol that controls the VCA's decay time.

Divide the LFO you're using to trigger the cymbal using a 4017, then use a divided output to control the vactrol. Whenever your LED turns on, you'll move the foot of the electronic drummer.

KICK DRUMS
- 250 kΩ potentiometer
- 2 x 100 kΩ potentiometers
- 4.7 kΩ resistor
- 10 kΩ resistor
- 3 x 1 MΩ resistors
- 2 x 4.7 nF capacitors
- 10 nF capacitor
- 2 x 100 nF capacitors
- 22 µF capacitor
- 40106 IC
- TL074 IC
- L7805 voltage regulator
- 2 x 1N4148 diodes

TOM-TOMS
- 250 kΩ potentiometer
- 2 x 100 kΩ potentiometers
- 33 kΩ resistor
- 10 kΩ resistor
- 4 x 2 MΩ resistors
- 2 x 1 nF capacitors
- 10 nF capacitor
- 2 x 100 nF capacitors
- 47 µF capacitor
- 40106 IC
- TL074 IC
- L7805 voltage regulator
- 2 x 1N4148 diodes

PROJECT: MAKING HARMONIC SOUNDS

Cymbals and other jangly things are unique in that they can be modeled inharmonically. But to mimic instruments with membranes—kick drums, tom-toms, and kettle drums—we'll need a circuit that can produce *harmonic* sounds. These are sounds in which overtones are evenly distributed, like in the spectrograms earlier in this chapter.

If you've ever held a live microphone up to a speaker, you've heard the screech of feedback. Feedback is what happens when you make a photocopy of a photocopy—the fine details slowly get lost while the weird nuances of the machine become more and more pronounced every iteration. Microphone feedback, believe it or not, is the same idea. The sound entering the microphone is played by the loudspeaker, picked up by the microphone, and reamplified again and again. This creates a runaway effect where our system spirals into chaos really quickly, but as soon as we distance the microphone from the speaker, we return to peace and quiet.

Normally, feedback is a thing you want to avoid, but this particular circuit makes clever use of it. The reason microphone feedback is so high pitched is because there's only a slight delay in the time between when the speaker plays a sound and the mic picks it up. But if we were to slow down the speed of sound, we'd be able to tune the frequency of the squeal. Couple that with amazingly fast hands, and we could play our microphone-loudspeaker combo like a synth drum. As ludicrous as it sounds, this discovery is at the heart of the drum machine.

We'll be constructing an active filter that's so resonant, any sudden change in voltage will shock the system into ringing feedback. By playing with the filter's cutoff, we'll be able to tune the feedback and change both its tone and its decay length. The circuit is surprisingly simple, as it consists of a high-pass filter and a low-pass filter bridged together at a single point. Because it looks like the letter *T*, this circuit is called a twin T.

Almost every electronic kick drum made before 1982 uses this circuit. The famous 808 kick drum is only a slightly more refined version. The CR-78 has scores of them—every rhythm attachment on the bottom of a tinny church keyboard too. This circuit is the bread and butter of electronic percussion.

KICK DRUMS

The TL074 operational amplifier (or op amp)—introduced in Chapter 8—is a clever little type of integrated circuit (Figure **A**). It can goose up a signal's current by comparing the live, moving voltage to a steady DC voltage. By outputting the *difference* in voltage, it amplifies the weak signal.

There is one more part to meet: the L7805 voltage regulator. It has three pins. One attaches to 9 V, one attaches to ground, and the third will provide a constant 5 V DC output. We'll need this regulator in order to provide a second voltage level to the circuit.

The twin-T filter is built around the op amp, with its outputs hooked up to its inputs. By tuning the potentiometers, we can make the circuit feed back—you'll hear the circuit start to oscillate! The schematic below shows the resistance values you'll need to make this circuit sound low and bass-y enough to emulate the low frequencies of a kick drum. The second potentiometer lets us toy with the envelope. By "choking" the feedback, we can deaden the ringing of the sound, turning a hum into a thud (Figure **B**).

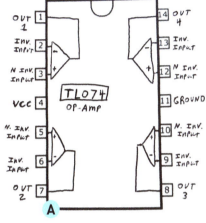

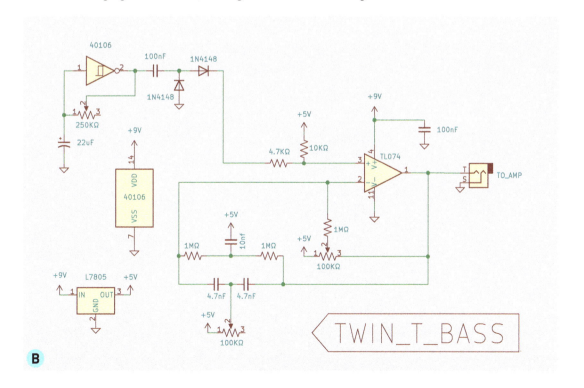

If you're having difficulty finding a sound, keep messing with the potentiometers. Not all combinations of resistances will produce a satisfying drum sound. In fact, for this very reason, most machines that employ a twin T don't even allow you to access these pots!

TOM-TOMS

Here's another twin tT with a very slight twist. This circuit is tuned to resemble a tom-tom rather than a kick drum. The component values are slightly different, but the theory is precisely the same. These guys sound particularly "drum machine-y":

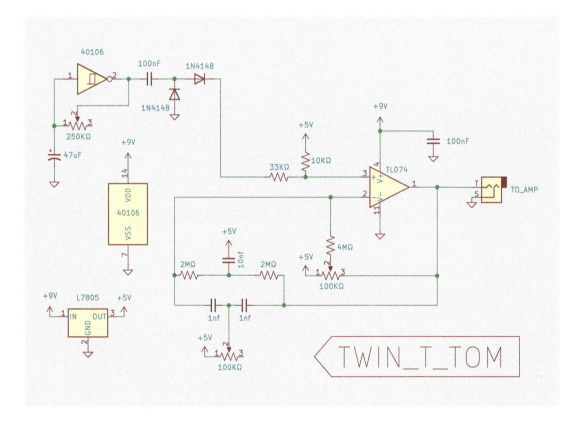

PROJECT: MAKING NOISE

Noise is a third type of sound that isn't really harmonic or inharmonic. White noise is a truly random signal—so random that *all* frequencies across the spectrum are audible. While inharmonic sounds might seem discordant, noise takes that to the extreme—resembling *hissing* more than anything else. Some people listen to white noise machines to help them sleep because the white noise helps camouflage any frequencies lingering in your bedspace (rumbling trains, snoring partners, monsters under the bed, etc.).

The method we'll use to make noise has its history in the 1920s. While researching resistors at Bell Labs, an experimental physicist with a delightfully redundant name, John Johnson[2], noticed they all exhibited some sort of low-level noise. This was a big problem for the engineers at Bell, who then had to spend decades figuring out how they could eliminate this "Johnson noise" from their circuitry. Amusingly, we'll be laughing in the faces of those engineers as we deliberately make a circuit that's *as noisy as possible*.

To make this offense to John Johnson and his buddies, we'll use a transistor—a BC548—and some good, old-fashioned hubris. By leaving the collector pin of the transistor disconnected, our transistor will amplify the Johnson noise and produce something that's quite snarelike.

SNARE DRUMS

To complete our drum kit, we can't get by without a snare. From an acoustics perspective, snares are unique. They combine the resonance of a membrane drum (which we simulated with a twin-T filter) with a burst of noise. If we cleverly combine these two, we can produce the *rat-a-tat-tat* you've come to know and love. See the schematic on the next page.

2. Sadly, not born in Wisconsin.

MATERIALS:
- 250 kΩ potentiometer
- 100 kΩ potentiometer
- 2 x 50 kΩ potentiometers
- 2.2 kΩ resistor
- 2 x 33 kΩ resistors
- 3 x 22 kΩ resistors
- 3 x 300 kΩ resistors
- 4 x 10 kΩ resistors
- 3 x 1 MΩ resistors
- 3 x 2 MΩ resistors
- 2 x 680 pF capacitors
- 4 x 1 nF capacitors
- 10 nF capacitor
- 5 x 100 nF capacitors
- 47 µF capacitor
- 2 x TL074 ICs
- BC548 transistor
- L7805 voltage regulator

Using a TL074 as a mixer, we can combine the two circuits and fine-tune their levels until the sound makes your face melt:

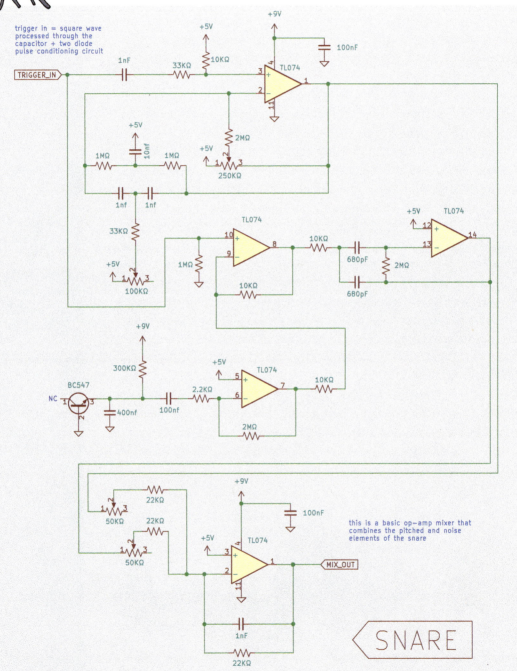

PROJECT: PERCUSSION SEQUENCING

There's plenty of fun to be had by playing the drum sounds individually, but the secret sauce of a drum machine is to have it *play for you*. These circuits can be hooked up to our sequencers from yester-chapter just as easily as our synth voices.

Historically, a lot of drum machines have come with preprogrammed rhythmic sequences—buttons with names of musical genres such as bossa nova or waltz printed on them. We can use a 4017 decade counter and have individual steps trigger drum sounds (with a little bit of fiddling). If you'd like a kick drum on step number 4, for instance, you can use a jumper wire to attach the pin associated with step count 4 on the 4017 to the trigger for your twin T. If you'd like a kick *and* a snare to hit on the fourth beat, add a second wire so that 4017 pulse triggers two sounds.

What we find particularly interesting about this process is not so much *how to sequence* but more *how a slight change in a rhythmic pattern can create a whole new canon of music*. For instance, changing which step of the beat pattern comes first is arbitrarily simple as far as engineering goes: all you have to do is shift all the wires over a step.

However, the cultural implications of such a small change cannot be understated. When you change which step is the downbeat of a pattern, for instance, you're actively taking a piece of culture and reprioritizing parts of it. By changing which beats in a rhythmic pattern have stress, you're inherently making a comment about the ways people's bodies should move. The joy of sequencing a drum machine, in my humble opinion, is not so much to program patterns that have existed for centuries but to deliberately mess with them and see what novelties result.

Unfortunately, creating a drum sequence is not as easy as merely plugging the respective outputs into the drum triggers. When multiple 4017 outputs are summed together, the active-high and active-low states cause conflict, and the output ends up being something in the middle of 0 and 1. In order to compose a drum sequence, we have to keep these two rules of thumb in mind:

1. For every 4017 trigger going to the same drum voice, send them through an OR gate so that you don't have two outputs fighting to determine the

MATERIALS:

BOMBA
- 500 kΩ potentiometer
- 2 x 2.2 kΩ resistors
- 2 x 100 nF capacitors
- 2 x 10 µF capacitors
- 40106 IC
- 4017 IC
- 4070 IC
- 10 x 1N4148 diodes

BOSSA NOVA
- 500 kΩ potentiometer
- 7 x 2.2 kΩ resistors
- 2 x 100 nF capacitors
- 10 µF capacitor
- 47 µF capacitor
- 40106 IC
- 2 x 4017 ICs
- 4070 IC
- 20 x 1N4148 diodes

logic level when one is output high and the other is active low (i.e. pulling down to ground). In our case, we use XOR gates because we have them and because, for our purposes, they are functionally equivalent.

2. For every consecutive trigger going to the same voice, you must build a diode AND gate, the inputs being the trigger step in question and the clock. (This is so the resulting output on two consecutive pulses isn't just solid on all the time.)

A SIMPLE EXAMPLE: BOMBA

A classic drum machine pattern is the *bomba sicá,* which originated in Puerto Rico. This particular rhythm is played on a drum called a *buleador*, which is alternately struck on top and hit on its side with a stick. These two voices—high and low—can be represented with this block notation:

If we want to sequence this rhythm, we'll first construct two electronic drums (whichever you fancy). We'll need the materials listed on page 245 to build a sequencer.

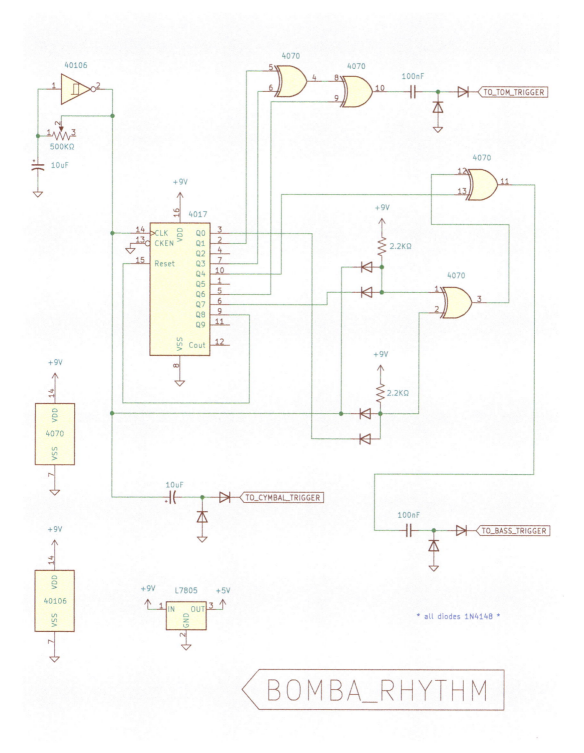

BOMBA_RHYTHM

A COMPLEX EXAMPLE: BOSSA NOVA

Pattern programming becomes even more complicated for longer patterns, such as a bossa nova (another staple drum machine preset):

The bossa nova pattern uses the same rules of thumb we learned with the *bomba* pattern. Each trigger going to the kick drum goes through an OR gate. Each trigger going to the snare also goes through an OR gate. Finally, consecutive triggers go through an AND gate with the master clock. The only additional step we've taken here is to double our sequencer stages from eight to sixteen. We can do this with the addition of another 4017, which we've set up to cascade, as seen in the schematic below.

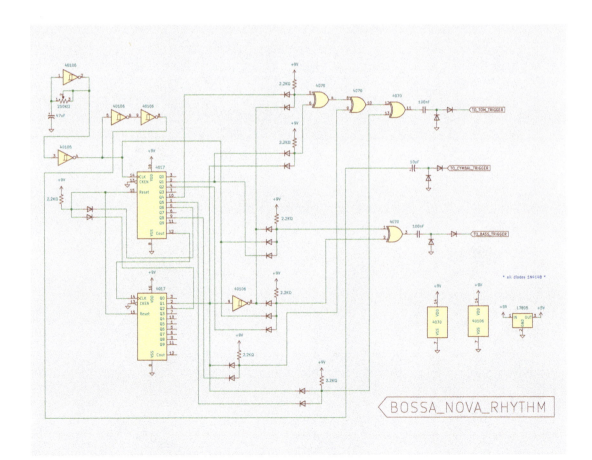

GUIDE TO RHYTHMIC MOTIFS

While corporations have saddled drum machines with a particular canon of
rhythms, there's no stopping you from making your own. Here is a series of our
select favorite rhythmic patterns:

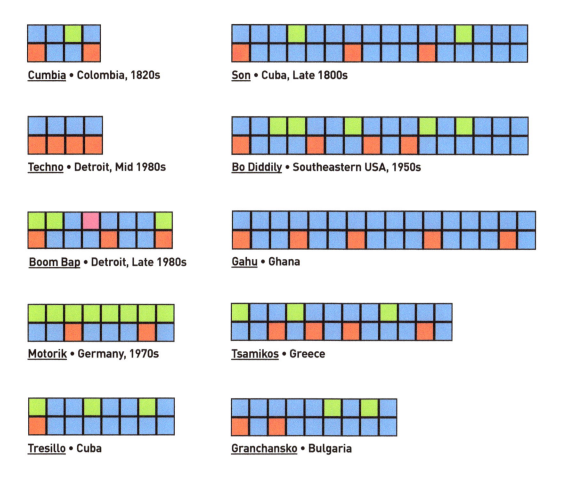

Cumbia • Colombia, 1820s

Son • Cuba, Late 1800s

Techno • Detroit, Mid 1980s

Bo Diddily • Southeastern USA, 1950s

Boom Bap • Detroit, Late 1980s

Gahu • Ghana

Motorik • Germany, 1970s

Tsamikos • Greece

Tresillo • Cuba

Granchansko • Bulgaria

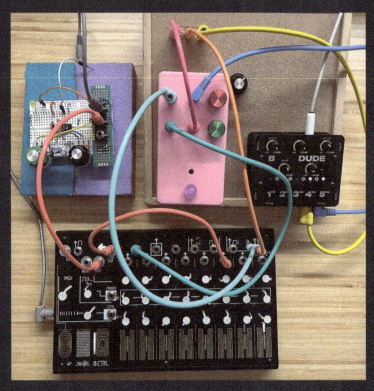

Musical Innovation: The Passive Voice

Artist: Jonathan Freitheim

About: From the artist: "A sound I like with breadboard synths is the moment you disconnect the battery. You might get a pop, a hiss, sometimes a little flutter or a *floop*. Unexpected! Uncontrollable! Fun! My experiments after that were dead simple: 'What happens if I replace the 9 V battery with control voltage from another synth I already own?'"

A way to start is to breadboard a super-basic 4093 oscillator circuit, make sure it sounds all right, and then remove the battery. Next, connect the cable tip from a CV source to plus and the cable sleeve to minus. Try it with an LFO, a spiky envelope, a pitch signal, triggers, gates, or who knows what else. Depending on the signal, you might brush against that magic threshold where the chip will start or stop functioning—that's where you want to be! These CMOS (complementary metal–oxide–semiconductor) chips are pretty hearty in that regard, so you can play around and see what works. Try it with other circuits that aren't too picky about voltage levels, and see what happens.

This pink box is just that, and I use it all the time with different synths and sequencers. The left jacks are connected to the plus and minus rails of two separate 4093-based square wave oscillators. The right jacks are audio out. You can even self-patch it into a quasi-Undertoner that sounds awesome with a regular old volt/octave CV pitch signal powering it. Is the result volt/octave pitch tracking? Absolutely not, but it is fun! Run that through a DIY low-pass gate, and, *blammo*, you got funky bass. Adjust those pots, and maybe you get funky squelchy bass. A slow LFO tweaked just right?—hissing screams. Use a fast-decay envelope as the power source, and you might hear finger snaps and hand claps.

13

Phase Locked Bass

There's one last IC we're going to discuss in this book, and it's a doozy. We've saved it for last, both to keep you in a state of unmitigated suspense and because it's *by far* the most complicated.

The IC du jour is the CD4046, known to her buddies as a phase-locked loop. This chip was developed for a variety of nifty mid-twentieth-century applications, including walkie-talkies, touch-tone phones, and all sorts of dangerous-looking radio equipment. A phase-locked loop works with two signals, and nudges one of them whenever it is out of sync with the other. Whenever two electronic signals need to be in synchronization, there's a phase-locked loop at the bottom of it. All broadcast transmission equipment runs on phase-locked loops.

To make the inner workings of the IC a little clearer, let's discuss what *phase* is. Imagine you're back in elementary school, and recall the day you brought in all your dinosaur toys. Not every kid is into dinosaurs at the same time, of course—unfortunately, you learn that most of your friends have passed that phase and have moved on to ninjas. You are excommunicated from the ninja lunch table and forced to sit with the less-cool dinosaur lovers in the back. "We're going through the same phase," your fellow dinosaur lovers tell you.

In signal processing, *phase* refers to *where* in a waveform one is. You and your ninjaphilic classmates are on the same general life path, but you're simply out of phase with one another.

For a more math-y example, look at the graph below. The red line is a sine wave, and the blue is a cosine wave. How do they compare?

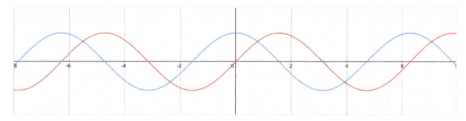

Sine waves and cosine waves, if you listen to them, sound identical. The only difference is that the cosine is pushed a quarter-period of the way to the right. Practically, this means a sine wave is identical to a cosine wave *if you delay its start by a few milliseconds*. We say the cosine wave is 90 degrees out of phase if you measure in radians. (A radian is equal to about 57 degrees.)

If you were to flip an audio signal on over its axis (or push it a half-period to the right), you'd create a wave that is 80 degrees out of phase. It's amusing to consider that this new, upside-down wave would sound the same as the original. However, as soon as you play the waves *together*, you'll find that they work against each other and cancel some parts of the sound out. If you play a

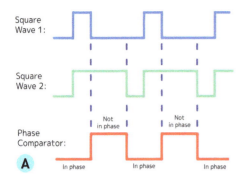

Square Wave 1:

Square Wave 2:

Phase Comparator:

A

Not in phase | Not in phase

In phase | In phase | In phase

SQ (square wave) 1 versus SQ 2 versus phase match

sound out of a single speaker at the same time as its 180-degrees-out-of-phase partner, they'll cancel each other perfectly, and all you'll hear is silence.

When dealing with square waves, however, this becomes even easier to think about. Because an ideal square wave has only two states—on and off—we can think about phase as whether or not the two waves share the same state.

Let's take two arbitrary square waves, as seen in Figure **A**.

Notice that when both square waves are on or off (i.e. they share the same state), they are *in phase*. When one is on and the other off, they are not in phase. A component called a phase comparator will output voltage when the inputs are *not in phase*. Eagle-eyed readers will notice this relationship is one we've already seen—it's just an XOR gate by another name.

Most phase-locked loops (called PLLs by those in a hurry) make use of a standard synthesizer component: a VCO, or voltage-controlled oscillator. There's actually a premade VCO inside the chip! This internal oscillator is what the PLL uses as a reference—it compares an incoming signal to this internal signal. Whenever the input square wave and the internal VCO are out of phase, the PLL emits current from one of its pins. This small burst of current can be hooked up to the VCO, causing it to speed up, nudging the signal just a tiny bit. This process continues over and over until both the internal and external signals match in phase and the nudging of the reference wave stops.

While PLLs weren't created with musical purposes in mind, they lend themselves oddly well to a lot of musical concepts. We can use the PLL to not only match pitch but also to do surprisingly complex math with rhythm. While it's not as immediately intuitive as, say, the 4040, this chip never stops inspiring once you get over the first few hurdles.

Here's a look at the inside of the 4046 (Figure **B**).

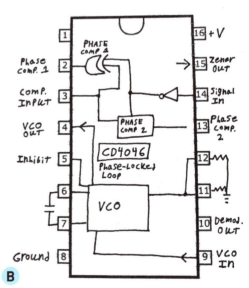

B

PROJECT: THE 4046 VCO

INSTRUCTIONS:

By far the easiest thing you can do with the CD4046 is create a VCO. The oscillators we've built so far have been *current* controlled—they all use a potentiometer to change resistance between two points, which changes the pitch. A VCO, on the other hand, needs only one input—if the potential energy at that input goes up, so does the frequency.

Amusingly, this is already done for you! The chip has a premanufactured VCO inside it, and all it takes to get it to sing is a potentiometer, a resistor, a capacitor, and a few wires. Take a look!

MATERIALS:

- 10 kΩ potentiometer
- 10 kΩ resistor
- 100 nF capacitor
- 4046 IC

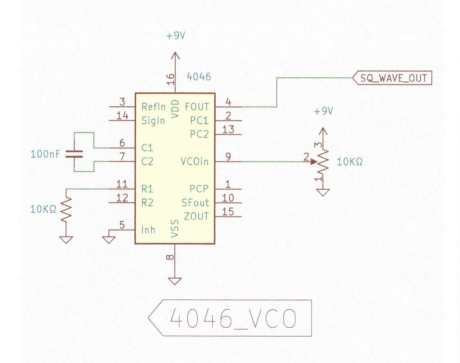

In order to control our VCO, we use a potentiometer as a *voltage divider*. We can make the world's simplest voltage divider from two passive resistors of the same value. Here's the schematic:

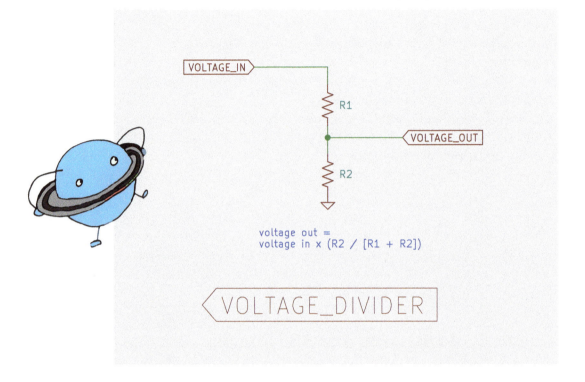

VOLTAGE_IN

R1

VOLTAGE_OUT

R2

voltage out =
voltage in x (R2 / [R1 + R2])

VOLTAGE_DIVIDER

If we know the input voltage and values of both resistors, we can calculate the output voltage with this simple formula:

Vout = Vin * (R2/(R1+R2))

For example, if we used two 100 k ohm resistors and a supply voltage (Vin) of 120 V, our output voltage would be (100 k/(200 k)) = 60 V. With just some middle school algebra (and a big bag of resistors), you can create any voltage (smaller than your input voltage) you'd like.

You'll notice that the potentiometer here isn't bridging two pins—it's being used to divide the voltage from your positive power rail. (We discussed voltage division in Chapter 8 on filters.) If you unplug the potentiometer and tap pin 9 while listening to the output, you'll likely hear some crazy noises. These noises are because the pin is quite sensitive to voltage changes, and your body affects the voltage going to the circuit.

PROJECT: PORTAMENTO, GLIDE, AND GLISSANDO

INSTRUCTIONS:

An interesting variation can be made to the VCO, and that's the addition of *glissandi*. You might call this *glide* if you're a synth person, *portamento* if you're a conservatory person, *slew* if you're an engineer, or *make the note bend slowly* if you're a normal person. Instead of immediately jumping from one note to another, this glissandi feature allows notes to slide from one to another—like how a violinist can move between notes without picking up their fingers. Best yet, you can adjust the amount of slide between notes.

This is quite simple to accomplish—by adding a capacitor to ground from the VCO input, we'll slow down how quickly the voltage can change. Adding a bigger capacitor will make the glide time longer.

In the following schematic, we've also added two more components: a tact switch and a pull-down resistor. When you hit the tact switch, you can trigger the oscillator to glide from the value it was previously playing to the new value chosen by the potentiometer.

MATERIALS:

- 10 kΩ potentiometer
- 10 kΩ resistor
- 470 kΩ resistor
- 100 nF capacitor
- 10 µF capacitor
- 4046 IC
- Tact switch

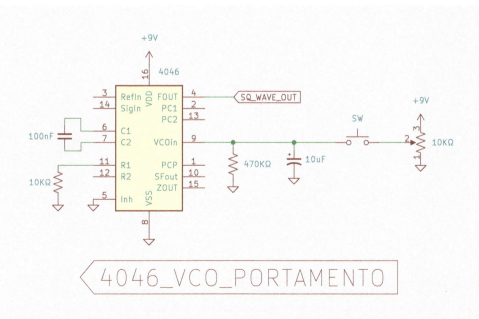

PROJECT: VCO SWITCHING

MATERIALS:

- 3 x 10 kΩ potentiometers
- 3 x 100 kΩ potentiometers
- 470 kΩ resistor
- 100 nF capacitor
- 1 uF capacitor
- 4.7 µF capacitor
- 10 µF capacitor
- 4046 IC
- 4066 IC
- 4093 IC
- 3 x tact switches

INSTRUCTIONS:

This simple but quite lovable circuit uses four oscillators—three LFOs from a CD4093, and one audio-rate VCO from a 4046. The three LFOs toggle on and off three gates on a 4066. Each of these gates lets through a voltage going to the VCO input. Because our voltage dividers are additive, we'll hear a different note come from the 4046 whenever a different combination of 4066 gates are open. The result is a lovely and somewhat chaotic jumble of notes. If you set the LFOs to slow and unrelated frequencies, you'll create an algorithmic composition that might take several dozen seconds to play and repeat.

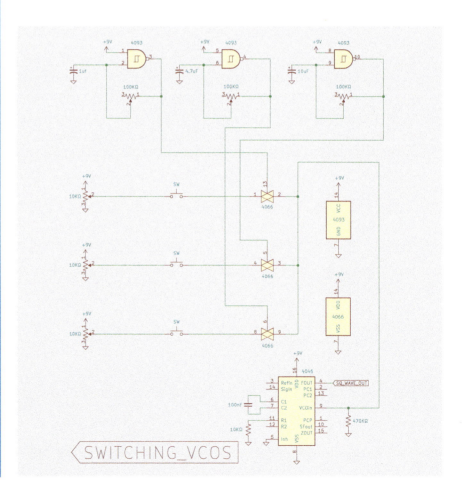

PROJECT: SAMPLE AND HOLD

INSTRUCTIONS:

Sample and hold is a tried-and-true synthesizer feature used in every science fiction movie. The result—typically—is a flurry of notes at seemingly random frequencies, which somehow became codified to mean "the sound a computer makes." Sample and hold is an incredibly uncreative name, but unlike most names for things in synthesis, it actually tells you what it does.

An input wave, often just white noise, is sent into a sample-and-hold (S+H, if you're in a hurry) circuit. At regularly repeating intervals, the S+H circuit takes a snapshot of the voltage of the input wave. The circuit then holds this instantaneous voltage level at the output until the next snapshot is taken. If you take snapshots eight to ten times per second, you'll hear an output that sounds rather R2-D2-ish. We can use the PLL to make an S+H circuit that while far from perfect, should scratch that sci-fi itch of yours.

If you recall, our 4093 can be used for both square and triangle waves. If we attach a slow-moving triangle wave to our 4046 input, we'll hear our VCO pitch oscillate up and down like a siren. By adding a large capacitor from the VCO input to ground, you'll notice that our VCO "remembers" its pitch, even after the triangle wave is disconnected. This is because the capacitor is holding that charge. (You will notice that the longer you wait, the more the pitch will start to bend downward.)

By putting a fast-gating 4066 switch between the triangle wave output and the VCO input, we can automate the process of manually disconnecting the two chips. If you set your potentiometers to the correct rates, you'll hear a sort of demented scale of notes going up and back down, following the contour of the triangle wave. See the schematic on the following page.

MATERIALS:

- 2 x 100 kΩ potentiometers
- 470 kΩ resistor
- 220 nF capacitor
- 1 µF capacitor
- 2 x 10 µF capacitors
- 4046 IC
- 4066 IC
- 4093 IC

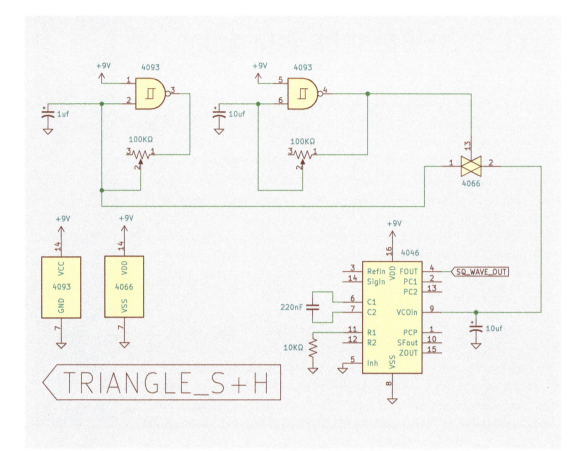

TRIANGLE_S+H

PROJECT: LFO MULTIPLICATION

INSTRUCTIONS:

You'll notice none of the previous circuits in this chapter really used the phase-locked loop for, well, anything except the VCO. Finally, we'll dig into a usage that actually involves phase comparison. Perhaps the coolest, and most complex, usage we have for the CD4046 is to use it to generate multiples of a frequency.

The applications of this circuit are myriad, but perhaps our favorite is using it to sequence triggers, either for electronic percussion or our envelope generator from the modulation chapter.

As we discussed before, our phase detector (at pin 13) looks at two square waves and outputs current when the waves are in different phases. If the phase detector starts outputting a lot of current, it means our waves are really out of sync and they need to be adjusted. What we really need is an *average* of what's coming out of pin 13.

We can create an average by using a low-pass filter—let's add a resistor and a capacitor to the comparator. Next, we'll connect the low-passed output to the input of the VCO. As more charge accumulates in the capacitor, the VCO slowly speeds up. As soon as the two signals are in phase, pin 13 stops outputting current, and the smoothed-out average voltage drives the VCO in to be the frequency it needs to be. This process happens incredibly quickly— so quickly that your two square waves can be synchronized in a fraction of a second.

In order to make this a multiplier, we can spice things up by adding one of our divider chips—a 4040 or a 4017. By using the VCO to clock the divider and comparing a *division* of the VCO to itself, we can create a multiple of our input frequency.

Let's illustrate this with an example. Our CD4046 compares what's going on at pin 14 and pin 3. If pin 14 is receiving 120 voltage signals a minute (120 beats per minute, or bpm), and pin 3 is receiving 70 bpm, pin 13 will react appropriately and send out a lot of tiny signals to indicate that the waves are not in sync. These tiny signals will tune the VCO until pin 3 matches 120 bpm. However, because the 120 bpm signal is coming from our divider—a division of a *much faster rate*, the result is a signal faster than your input.

Yes, we know this sounds complicated, but it really does work. Your PLL is determining how much time elapses between oscillations and rapidly

MATERIALS:

- 100 kΩ potentiometer
- 250 kΩ potentiometer
- 470 kΩ resistor
- 47 kΩ resistor
- 330 Ω resistor
- 1 µF capacitor
- 4.7 µF capacitor
- 2 x 10 µF capacitors
- LED
- 4046 IC
- 4017 IC

tuning itself to a division of those oscillations determined by the ratios in your 4040 or 4017 chip. If you take a second to think about it, you'll realize this is tremendously cool.

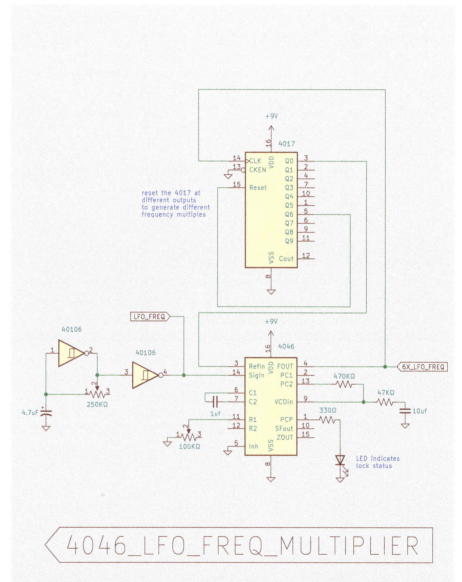

PROJECT: AUDIO-FREQUENCY MULTIPLICATION

MATERIALS:
- 10 kΩ potentiometer
- 250 kΩ potentiometer
- 470 kΩ resistor
- 47 kΩ resistor
- 330 Ω resistor
- 100 nF capacitor
- 0.1 µF capacitor
- 1 µF capacitor
- LED
- 4046 IC
- 4017 IC

We can use the same general idea in LFO multiplication for multiplying audio frequencies as well, just with a bit more finagling. We'll switch out the two capacitors and the potentiometer connected to the 4046 for slightly faster frequencies.

Each of the 4017's outputs can be connected to the reset pin for different harmonies, as seen in the schematic.

If you'd really like a crazy variant, pass each of the 4017's outputs to the branches on a 4051 multiplexer. You'll be able to sequence the harmonies to your heart's content. ♪

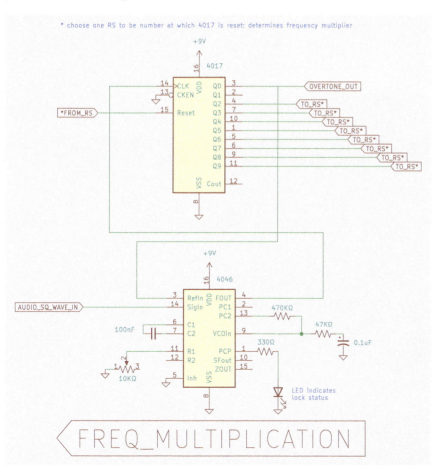

Musical Innovation: Heat Sink Resonator

Artist: Kel Smith

About: From the artist: "I've always been fascinated by the relationship between machines and human capability. I started Suss Müsik in 2016 as a vehicle to explore these obsessions in creative form. Some devices are built from archaic consumer technologies (like 1990s-era hard drive enclosures),while others are custom designed and manufactured via 3D printing or other methods. Sometimes, I'm able to reuse circuitry from the electrical boards I recycle—potentiometers and capacitors tend to work best—but it can be tricky, and results are often spotty. Mostly, I reconstitute old boards into new enclosures for MIDI devices and rudimentary synths.

"I started with simple oscillators built from a 555 integrated circuit, but my go-to chip these days is either a CD4093 or something customized for the Arduino Micro. One of my pieces derives sound from two piezo microphones attached to a computer heat sink. A humming resonance can be achieved by simply running my hand over the heat sink's 'ribs,' while striking it softly with a mallet produces a lovely ringing tone. The actual output is powered by an LM386 chip, which besides serving as an oscillator makes for a nice little amplifier."

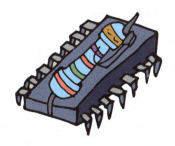

Part IV

PUTTING IT ALL TOGETHER

Congratulations, you brilliant engineer you. You've made it to the final section! Now what?

The greatest part about breadboarding circuits is that it's easy to integrate them into one another. In fact, pretty much all the projects in this book can work alongside one another seamlessly. When we teach workshops, we encourage people to pick their favorite generators, modifiers, and outputs and combine them into an exciting chimera of an instrument. For instance, you could put a square wave into a filter and out of a talkbox. Or you could make a suite of speakers that plays the 4017 wavetable circuit. The possibilities are endless!

Of course, we weren't really satisfied leaving the book at a choose-your-own-adventure conclusion. So for those who would really like a challenge, we've cooked up something especially for you.

I n an effort to graduate you from your first year of Dogbotic Labs, we've decided to gift you a truly ludicrous project that we're calling the Dogbotophone MK1. The Dogbotophone isn't anything new—in fact, it's really just a reprise of some of this book's more enigmatic circuits. However, this circuit provides a way for you to combine a full orchestra of instruments together in one unified, playable interface. The Dogbotophone MK1 boasts a full percussion section capable of playing virtually any sixteen-step sequence you could dream of, a melodic sequencer that composes for itself, a bass sequencer that works in counterpoint, an active filtering system, and an active voice mixer. The entire circuit can be rewired in real time, making it legitimately performable.

As is the case with many large circuits, the Dogbotophone MK1 cannot fit on a single page of this book—and even if it could, it would be really small and annoying to read. Instead, we've divided the circuit into fifteen bite-sized pieces and shown you how they connect to each other. We recommend building the circuit step by step, and checking to make sure each step works properly before advancing.

AN OVERVIEW

This simple block diagram shows you where we'll be headed on our little odyssey. At its core, a clock LFO produces slow-moving square waves, which will be used as our instrument's internal tempo system. The clock LFO provides time information to three voice modules: the overtone sequencer, which produces melodies; the undertone sequencer, which produces a bass counterpoint; and the drum sequencer, which will give us some delicious electronic percussion accompaniment.

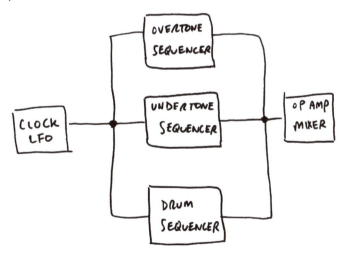

POWER AND VOLTAGE REGULATION

The below schematic shows the supply voltage pins for the circuit's IC units.

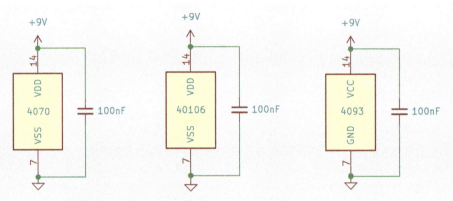

These are the supply voltage pins for the logic ICs used in the circuit. Note that for other ICs (e.g. op amps) the convention is to depict the supply voltages on the first unit of the chip.

The capacitors above are called "bypass capacitors" — they improve the noise immunity of the circuit by stabilizing the supply voltage at the ICs.

Below is a voltage regulator, used to turn the +9V of your circuit supply into +5V.

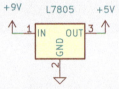

LOGIC IC POWER PINS + VOLTAGE REGULATOR

Step 1: An Audio-Rate Oscillator

This simple 4093-based oscillator produces a square-wave tempo for the rest of the circuit. Moving the potentiometer or changing the capacitor will result in a new tempo (Figure **A**).

Step 2: A Sequencer Clock

Let's expand our oscillator with a 4040 binary divider. An input square wave goes from our oscillator into pin 10 of the divider and sends out three octaves of square waves through pins 9, 7, and 6. These three signals are really three digits of binary code that will be used to control three 4051 multiplexers that make up the Dogbotophone's three sequencers (Figure **B**).

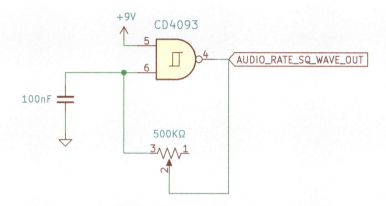

This circuit generates an audio-rate square wave used to clock the 4017s that drive the over- and under-tone synths of the circuit.

AUDIO RATE OSCILLATOR

A

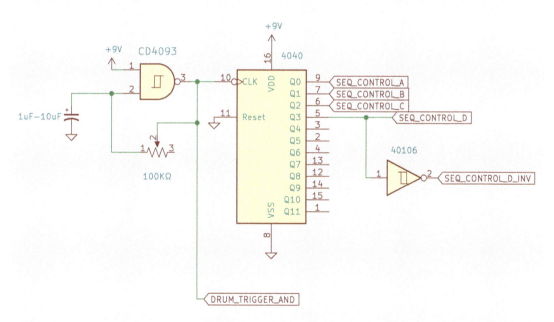

SEQUENCER CLOCK

This circuit generates binary code capable of addressing all the 4051 multiplexer chips that serve as the three different 16-step sequencers in the circuit.

B

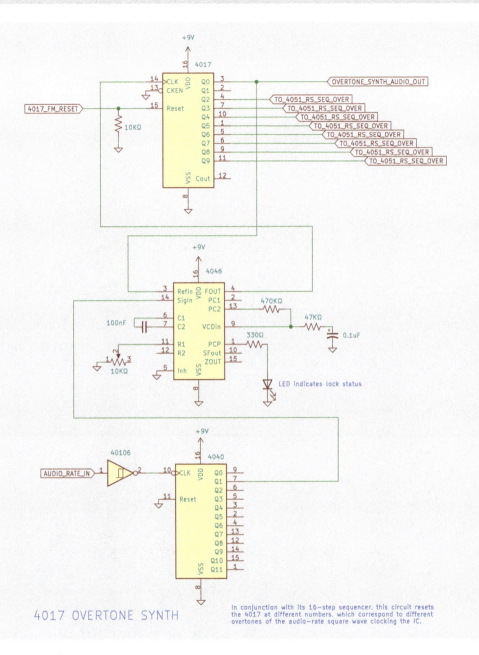

4017 OVERTONE SYNTH

In conjunction with its 16-step sequencer, this circuit resets the 4017 at different numbers, which correspond to different overtones of the audio-rate square wave clocking the IC.

C

Step 3: A 4017 Overtone Synth

This circuit is slightly more complex—but only slightly! Honest!

The 4017 Overtone Synth produces the overtones of the audio rate signal going to a 4040. Alongside a sequencer, this little synth voice can chirp out melodies that can't play "out of tune." (Figure **C**)

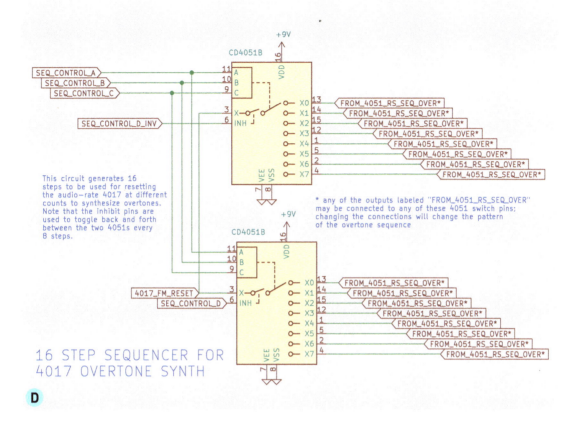

This circuit generates 16 steps to be used for resetting the audio-rate 4017 at different counts to synthesize overtones. Note that the inhibit pins are used to toggle back and forth between the two 4051s every 8 steps.

* any of the outputs labeled "FROM_4051_RS_SEQ_OVER" may be connected to any of these 4051 switch pins; changing the connections will change the pattern of the overtone sequence

16 STEP SEQUENCER FOR
4017 OVERTONE SYNTH

D

Step 4: A Sixteen-Step Sequencer for Overtone Synth

This sequencer takes our three-octave binary control (from step 2) and uses it to switch between sixteen steps of a melodic sequence. The sequence itself comes from the 4017 Overtone Synth from Step 3. By plugging in different combinations of wires, we can produce a practically unlimited number of sequences. Thanks to the 4017's harmonic resetting capabilities, the sequencer can be improvised with ease—it will never play a wrong note (Figure **D**).

Steps 5 and 6: A 4017 Undertone Synth and a Sixteen-Step Sequencer for Undertone Synth

The 4017 Undertone Synth is easy to build but hard to wrap your head around. The 4017 serves as a counter that fires off current from eight outputs. The count is reset every time one of those outputs fires current *if and only if* another sequencer happens to share the same step number. Alas, this is almost impossible to understand unless you look at the schematics (Figure **E** on the following page).

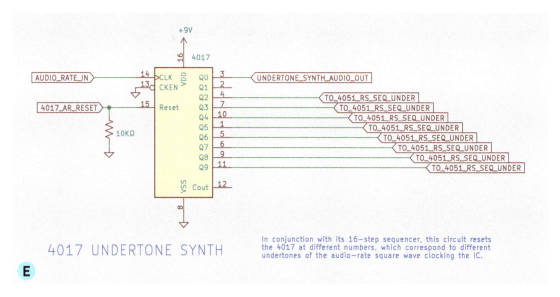

4017 UNDERTONE SYNTH

In conjunction with its 16-step sequencer, this circuit resets the 4017 at different numbers, which correspond to different undertones of the audio-rate square wave clocking the IC.

E

The beauty (and complexity) of this particular instrument is that it's built from two sequencers running at audio rate. When the two sequences happen to bump into each other, our speaker wiggles. The incredibly fast speed of the sequencers means that these infrequent bumps are still occurring at audio rate (Figure **F**).

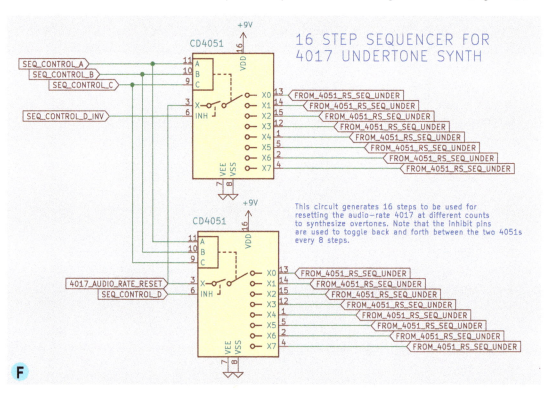

16 STEP SEQUENCER FOR 4017 UNDERTONE SYNTH

This circuit generates 16 steps to be used for resetting the audio-rate 4017 at different counts to synthesize overtones. Note that the inhibit pins are used to toggle back and forth between the two 4051s every 8 steps.

F

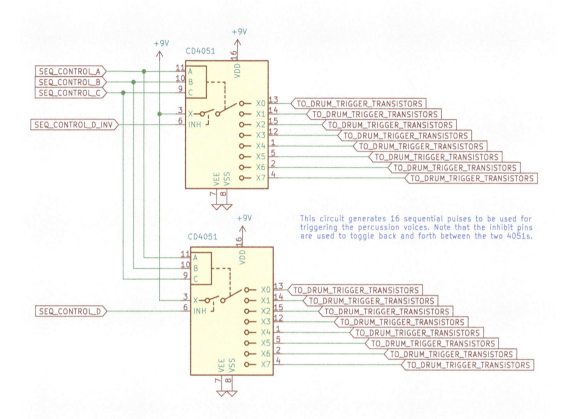

This circuit generates 16 sequential pulses to be used for triggering the percussion voices. Note that the inhibit pins are used to toggle back and forth between the two 4051s.

16 STEP SEQUENCER FOR DRUMS

G

Step 7: A Sixteen-Step Sequencer for Drums

In comparison to the last few steps, our drum sequencer is pretty easy to follow. Our clock from Step 2 controls sixteen steps of triggers that can control our drum voices (Figure G).

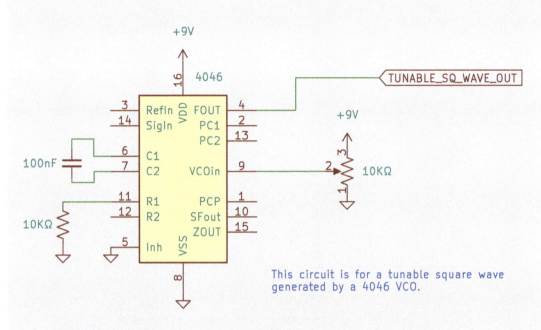

+9V

16 4046

3   RefIn   FOUT 4   TUNABLE_SQ_WAVE_OUT
14 SigIn   PC1   2
                  PC2   13   +9V

6   C1
100nF   7   C2   VCOin 9   2   10KΩ

11 R1   PCP   1
12 R2   SFout 10
10KΩ   ZOUT 15

5   Inh

This circuit is for a tunable square wave
generated by a 4046 VCO.

TUNABLE SQUARE WAVE

H

Step 8: A Tunable Square Wave
In addition to two sequenced lines, the Dogbotophone features a playable monosynth in the form of a 4046 square-wave oscillator (Figure **H**).

Steps 9 and 10: Twin-T Voices
The Dogbotophone has three drum voices: two harmonic and one inharmonic. The two twin-T voices can be tuned from a kick drum to a tom to a woodblock (Figures **I** and **J**).

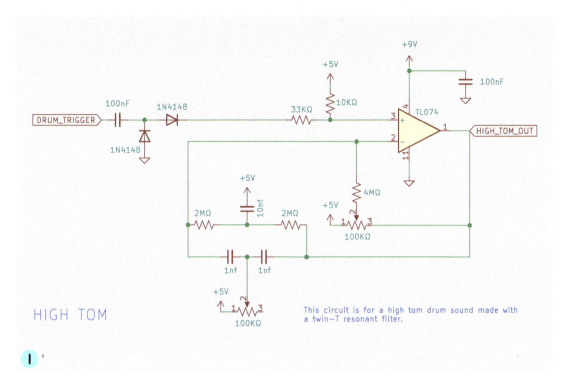

HIGH TOM

This circuit is for a high tom drum sound made with a twin-T resonant filter.

I

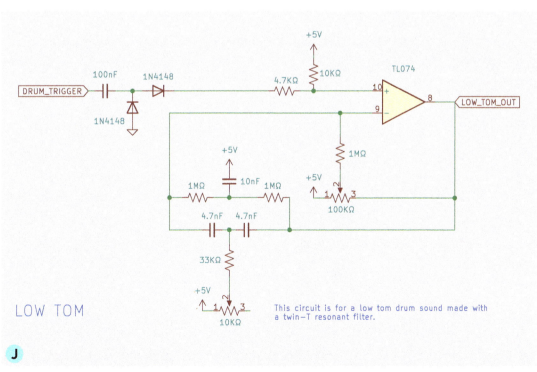

LOW TOM

This circuit is for a low tom drum sound made with a twin-T resonant filter.

J

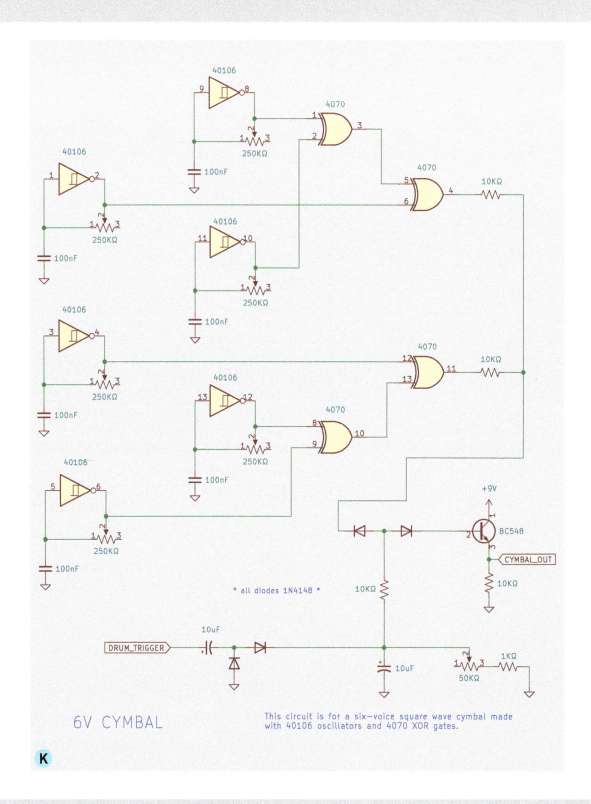

6V CYMBAL

This circuit is for a six—voice square wave cymbal made with 40106 oscillators and 4070 XOR gates.

Step 11: A Cymbal (XOR Voice)

A gift to you! This is another schematic that looks worse than it really is. This XOR gate cymbal uses six square wave oscillators (on one 40106), four XOR gates (on one 4070), and one diode-based VCA for a satisfying nostalgia trip (Figure K).

Step 12: Diode AND-Gate Sequencer Reset

An unfortunate aspect of the drum circuitry is that steps can't be added directly from the sequencer to the triggers. We'll have to add a reset switch. This small circuit can take any number of steps from the drum triggers and will trigger a drum when a control signal arrives coincident with a clock signal (Figure L).

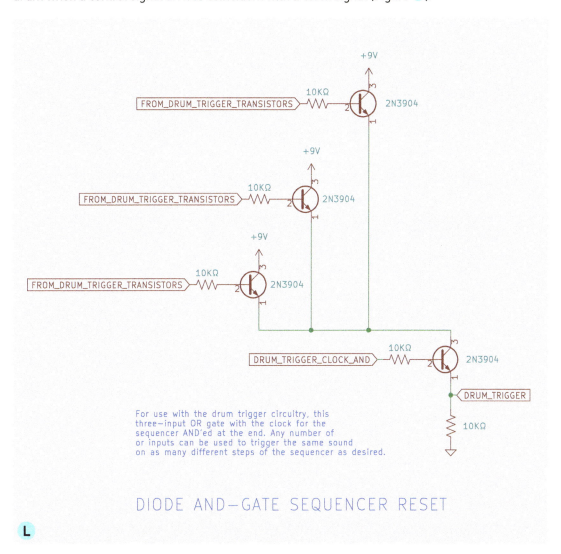

For use with the drum trigger circuitry, this three-input OR gate with the clock for the sequencer AND'ed at the end. Any number of or inputs can be used to trigger the same sound on as many different steps of the sequencer as desired.

DIODE AND-GATE SEQUENCER RESET

L

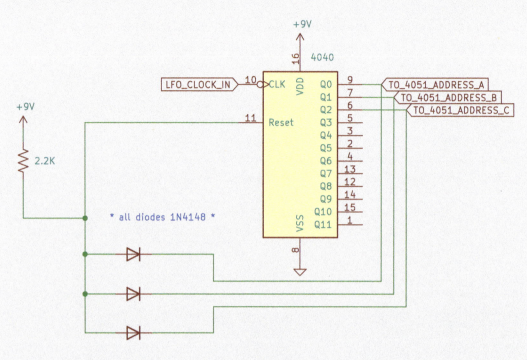

This is an optional circuit for changing any sequencer pattern
length with a diode AND gate to reset a 4040 clocking a 4051.

DIODE AND—GATE SEQUENCER RESET

M

Step 13 (Optional): Global Sequencer Reset

If you'd like to change the length of the sequencer patterns on the fly, consider
this small circuit. Resetting the 4040 can be done by running three binary digits
of your choosing through an AND gate (Figure **M**).

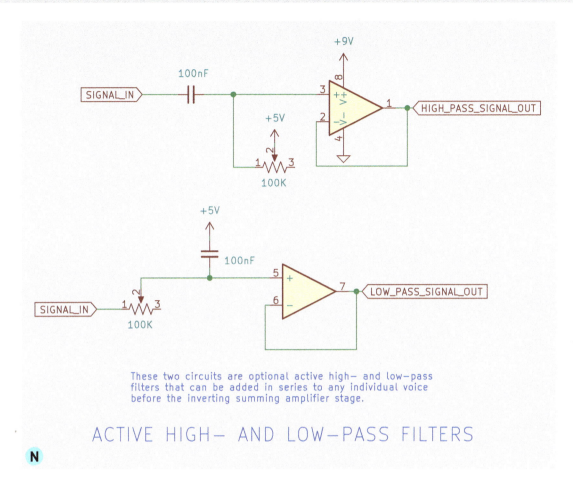

SIGNAL_IN
100nF
+9V
3 ++ V+ 8
2 -\> V- 4
1
HIGH_PASS_SIGNAL_OUT
+5V
1 100K 3

+5V
100nF
SIGNAL_IN
1 100K 3
5 +
6 −
7
LOW_PASS_SIGNAL_OUT

These two circuits are optional active high- and low-pass
filters that can be added in series to any individual voice
before the inverting summing amplifier stage.

ACTIVE HIGH- AND LOW-PASS FILTERS

N

Step 14: (Optional) Active High-Pass and Low-Pass Filters

Any individual voices can be sent through a playable active filter before its mixing
stage. These are very simple circuit-wise but add a whole new dimension of
playability (Figure **N**).

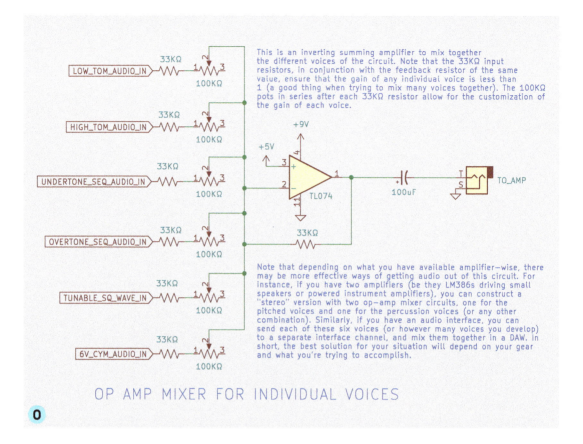

This is an inverting summing amplifier to mix together the different voices of the circuit. Note that the 33KΩ input resistors, in conjunction with the feedback resistor of the same value, ensure that the gain of any individual voice is less than 1 (a good thing when trying to mix many voices together). The 100KΩ pots in series after each 33KΩ resistor allow for the customization of the gain of each voice.

Note that depending on what you have available amplifier—wise, there may be more effective ways of getting audio out of this circuit. For instance, if you have two amplifiers (be they LM386s driving small speakers or powered instrument amplifiers), you can construct a "stereo" version with two op—amp mixer circuits, one for the pitched voices and one for the percussion voices (or any other combination). Similarly, if you have an audio interface, you can send each of these six voices (or however many voices you develop) to a separate interface channel, and mix them together in a DAW. In short, the best solution for your situation will depend on your gear and what you're trying to accomplish.

OP AMP MIXER FOR INDIVIDUAL VOICES

0

Step 15: An Op-Amp Mixer

Finally, the op-amp mixer inputs six voices and lets us mix them to taste (Figure **0**):

- **Three drum voices**
  - Our two twin-T circuits
  - Our inharmonic cymbal circuit
- **Two sequenced voices**
  - Our overtone synthesizer
  - Our undertone synthesizer
- **One playable voice (our tunable square wave)**

If you'd like a stereo output, you could either build two TL074-based mixers and feed them individual signals or use our panning technique from our chapter about amplifiers (see page 78).

If you've made it this far, congratulations! Your room is likely a mess of wires, but you've *done it.* Now, to solder it all together! 🎵

15

THOUGHTS ON AUTOMATION

A s a part of promoting this book, my coworker Sean Hallowell and I spent a week at the Bay Area Maker Faire. Among our exhibits was an impressive modular drum machine that Sean had built from the circuits found in the chapter on drums. On the first day, one person approached us and, with total earnestness, told us that his daughter was a drummer and that we should feel ashamed for taking potential jobs away from her.

That interaction inspired us to write this chapter.

There was an era in recent history when people duly believed synthesizers were going to take their jobs away. Based on our experience at the Maker Faire—in a community filled with people obsessed with automation—many people still believe this.

The basic "synthesizers are bad" argument goes like this: As our technological prowess gets more and more advanced, synthesizers will sound increasingly indistinguishable from acoustic instruments. In a future where a synthesizer can replicate the sound of a horn section, it's more cost-effective to hire a machine than a five-person brass band. These arguments position the synthesizer not so much as a musical instrument but as a cost-cutting tool—a thing that mockingly reminds people *that music exists*. By playing a synthesizer, you're effectively telling all your hardworking musician friends that they don't matter.

This perspective definitely sounds a little dated today, but I heard this argument *a lot* growing up. (For context, I was born in 1994.) Most of the musical communities I was a part of—in the mid-2000s, mind you—would speak of electronic music in hushed tones. Listening to synthesizer music, let alone *playing* it, was seen as buying into some crass commercial mockery of more "serious" music. Electronic instruments were second rate compared to human-played instruments.

Being a synth-curious kid was certainly an act of rebellion, but it wasn't really a *cool one*. Movies of that era portrayed guitar players as suave, technically dexterous, and *dripping with sex appeal I can assure you they didn't have.* Every portrayal of a synth or drum machine inevitably involved it being played by a klutzy, talentless dork. It was a bad era to be a synth enthusiast.

What these communities around me didn't know was that synthesizers were *never* intended to be replacements for anything. They started out as odd curiosities for musician-engineers, and remained that way for decades. As far as I can tell, the antisynthesizer neurosis didn't set in until the early 1940s, when tinkerer Georges Jenny started selling the first synthesizer that "emulated" more traditional musical instruments. Although only a few hundred units were produced, Jenny's invention caused an existential crisis among music unions.

His synth, called the Ondioline, looks and sounds incredibly tame by today's

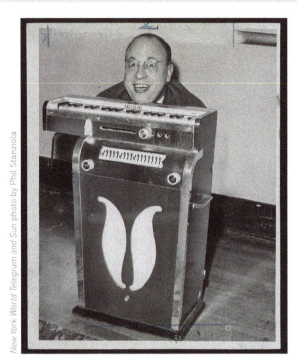

New York World Telegram and Sun photo by Phil Stanziola

Jean-Jacques Perrey and his Ondioline

THE LEADER OF THE LUDDITES

standards. But ironically, the Ondioline was a *complex* machine to operate—making it far more expensive to hire than the instrument it was "replacing." Every parameter had to be changed by hand, requiring separate manual controls for envelope, modulation, tone, and wave shape (see the bank of switches in the photo)—on top of individually playing the notes on a keyboard interface. Looking back on videos of Ondioline players from the 1950s, there's no doubt in my mind that the machine required incredible technical prowess to operate. To me, at least, it's every bit as impressive as actually playing the instrument it emulated.

Central to all these fears is the idea that the job of "musician" is an unchanging one. If a musician's financial situation depends on people needing a live clarinetist, it stands to reason that an automated clarinetist would be an economic force worth fearing. People's fear of electronic music stems from people's fear of capitalism. It's truly unnerving to think that your skills can be replaced, and for some reason this becomes much more nerve-wracking to people when we frame it in terms of cultural skills. After all, industrial machines have been replacing blue-collar jobs for centuries with little to no existential uproar.

There's one anecdote about replacement I find particularly telling nowadays. In the late 1800s, English textile mills started shifting their operations to favor machines as a means to cut costs. One textile worker, Ned Ludd, decided to protest these changes out of the fear they would ruin the livelihood for thousands of workers. He founded a political faction of workers that loudly and vocally opposed these changes, sometimes sabotaging the fancy new equipment in protest. The group took their name after their leader, and called themselves the Luddites. Isn't it odd how nowadays, a Luddite is a person who stubbornly resists technological change and not a job-insecure person concerned for the well-being of their family? Consider what that says about who gets to write history.

It's quite obvious when we look at the past few hundred years of history to note that the ruling class has almost always sided with automation. It's not totally unfounded to be anxious about electronic music—if people don't care whether their clothes are machine made, will they care if their music is? How far can we remove humans from the musical equation before our very definitions of music break down?

With very few exceptions, synthesizers were never intended to compete with acoustic instruments. Earnestly, thinking of synths as machines that emulate other machines drastically undersells what they're capable of. When you can make any sound in the world, why bother making a sound that already exists? It's as if you had access to a machine that could invent new cuisines but you used it exclusively for unsalted popcorn. A synthesizer's sound isn't limited by the stodgy laws of acoustics—in this book, we actively encourage you to relish in your circuit's eccentricities.

Today, the notion that human musicians will be replaced by human synthesizer players seems completely trite. Obviously, that moment never really happened— live music is as strong as it ever was. Today, the debate has shifted from "Will synthesizers replace instruments?" to "Will human musicians be replaced by computer musicians?"

As we are writing this book—in the mid-2020s—the landscape of computer-made art is shifting dramatically. Tools branded as generative artificial intelligence are churning out image after derivative image. Meanwhile, the folks who made the art that made up the initial dataset for generative AI are now losing work opportunities only to be replaced by a computer that can mimic their style. The same is happening with popular music. The shocking thing is that, by and large, *the public doesn't seem to care how the art is made.*

It's tempting to say that the corporate AI landscape is poised to change what an artist is. Perhaps in the near future, an artist will be defined not by the content they can produce but by their ability to distinguish "good AI content" from "bad AI content." While this might not sound so different from the anxiety Jenny's Ondioline wrought, the sheer volume of *stuff* that has been developed with AI has made it a lot harder to sift through the noise. Generative AI is a tool designed for legitimizing plagiarism that extrudes cultural by-products like a tube squeezes toothpaste. As of this writing, it is too early to tell if this will be yet another

Ondioline or something more sinister. No matter what I write, I'm most certainly going to be wrong.

It's important to be cautious of, but not to fear, tools. People also once believed that due to the incredible speed of locomotives, riding on one would induce madness—an argument you'd be hard pressed to hear nowadays. However, it's important to be cautious about how people in positions of power plan to use those tools. I don't see a future in which AI destroys culture—culture will always survive. In fact, I'm quite hopeful that AI could greatly reduce the amount of unnecessary toil and rote work that people have to do. But alongside that, I easily see a future in which AI is used to further consolidate money for corporations and universally worsen the wealth gap across our planet.

Our hope with this book is to encourage readers to look at sound synthesis as a living art—and that the mere process of experimentation puts *you* at the forefront of synth development. The fact is that synthesizers really don't "improve," at least from an artistic standpoint.

We're generally taught to think that the slow march of time indicates *improvement*. Many people seem to believe that music slowly changes over time, as if they're heading toward some sort of perfect form. They see the twentieth century's history of blues, jazz, rock, and rap as exhibiting some sort of trend in complexity. These people are wrong. Art has always been a response to the context in which it was created, which makes all expressions equally valid. The music of Katy Perry is no more or less complex than the music of J. S. Bach. They were simply made for incredibly different contexts. Take a close look at any piece of art and you'll find there are more than enough intricacies to talk about.

We opened this book with a discussion of the phrase "music technology" and how it dangerously prioritizes technique over art. It is not technical complexity but human context that determines the importance of a piece of art. As such, experimentation is always at the cutting edge of art—your experimenting with CMOS chips is every bit as valid musical research as anything else being done in the fanciest of laboratories. Ultimately, musical research involves determining which sounds are the most culturally interesting to you.

Building an analog synthesizer in the mid-twenty-first century is certainly anachronistic, but we've found it's a great way to make people consider their own artistic practice and the ways in which it lets them get under the hood. Corporations that fabricate products would like us to believe their goods are *mythically complicated*. It makes them look more impressive, and it ensures that you won't dare open it up and try to fix it. Everything designed comes with an established technological code—an ideology about how it should be used. This code, however, might not serve your needs.

In a world designed for you to buy and not to reuse, learning to build and maintain electronics is a very political act. While a lot of electronic music has been characterized as cold and unfeeling, the process of actually building these instruments shows you how beautifully lifelike and organic they can be, down to the electron level. After a bit of self-discovery, you might realize that electronic music isn't cold and unfeeling at all.

At the end of the day, it's important to remember that electronic music really isn't music made for machines. It's made for you.

Electronic music is something deeply, deeply human. 🎵

APPENDIX I

THE COMPONENTS OF A CLASSICAL SYNTHESIZER

While every synthesizer looks a little bit different, they mostly all work the exact same way. If you can understand what the different parts of a synthesizer do, you'll be well on your way to playing 90% of synthesizers in the world (or, at least, getting a sound out of them).

This chapter isn't exhaustive. People are building new types of synthesizers all the time that do any number of things. However, the parts below are ones you're likely to come across no matter which instruments you're looking at. If you see a synthesizer that doesn't have any of these components, it's trying to make a point.

Audio Generators

Audio generators are circuits that vibrate at frequencies that are high enough for humans to hear.

OSCILLATORS

Oscillators are generally the heart of the synthesizer, as they're the ones that do all the vibrating. An electronic oscillator is a circuit that moves between two voltage levels in a chosen pattern—typically sine, square, sawtooth, or triangle. A VCO (voltage-controlled oscillator) will change its pitch based on the voltage of a control signal. This control signal can be sent from a keyboard, a sequencer, a potentiometer that you twiddle, a photoresistor that you wave your hands over, or any number of other things.

To build a simple oscillator, see Chapter 3. To build a number of oscillators that integrate together in amusing ways, see Chapter 6. To build a VCO, see Chapter 13.

NOISE GENERATORS

Unlike an oscillator, a noise generator doesn't vibrate with any sort of clean pattern. Noise generators are great for creating loud snare drum sounds and, counterintuitively, noises that help people fall asleep.

To build a noise generator (to make a snare drum sound), see Chapter 12.

Amplifiers

Audio generators don't typically start and stop, like musical notes do. For that, we need an amplifier to turn down the signal when we *don't* want to hear it, and turn it up when we do. All amplifiers take a signal from an audio generator and beef it up—they use a small voltage signal to control a much larger voltage signal.

VCAS

A VCA (voltage-controlled amplifier) increases the amplitude of a sound proportional to the control signal it is sent. A higher control voltage means a louder signal.

To build a simple VCA (with decay control), see Chapter 10.

ACTIVE MIXERS

Synthesis gets a lot more interesting when you start to combine different signals. An active mixer is really a series of amplifiers that transforms any number of signals into a clean, unified mix. Without an active mixer, signals might be weak and/or suffer from cross-oscillator modulation.

To build an active mixer with op amps, see Step 15 of Chapter 14.

POWER AMPS

If you want to control a speaker with a voltage signal, a power amp is necessary. It's what transforms a low-current signal into a current strong enough to push and pull the magnet in your speaker. If you try to use a VCA or a line-level amplifier to power a speaker, you'll get a whole lot of nothing.

To build a simple power amp, see Chapter 4.

MODIFIERS

A modifier is an effect—a costume that your generator wears. Send an audio signal through a modifier, and you'll hear that it sounds different when it comes out.

FILTERS

An audio filter pulls frequencies out of an audio signal. We can use filters to change the harmonic spectra of sound. Most filters you can control have a *cutoff*, which signifies the frequency at which the filter starts cutting out frequencies. A low-pass filter cuts out the high frequencies (above the cutoff), allowing the low frequencies to pass through. A high-pass filter cuts out the low frequencies (below the cutoff), allowing the high frequencies to pass through. A band-pass filter cuts out both high and low frequencies (between two cutoffs), allowing a narrow band of frequencies to pass through.

To build all sorts of filters, see Chapter 8.

WAVESHAPERS

Changing the shape of a generator's wave is a great way to provide timbral variation to a sound. There are many ways to shape the wave of a sound, such as pulse-width modulation (PWM) and wavetable modulation.

To build PWM control, see Chapter 8. To build a 10-stage wavetable doohickey, see Chapter 9.

EFFECTS CHAINS

Some synthesizers come with their own prebuilt effects: distortion, delay, reverb, and so forth.

For a few effects, including an electromechanical reverb and a talkbox, see Chapter 4.

MODULATION SOURCES

Modulators are third hands that can change parameters for you automatically. Don't stand there twiddling your filter up and down—just use a modulation source!

LOW-FREQUENCY OSCILLATORS

A low-frequency oscillator is an oscillator that moves *really, really slowly*. We can't hear an LFO (it's too low pitched), but we can use it to modulate other parameters. For example, an LFO turning on and off an amplifier could be used as a tremolo effect. An LFO turning up and down the frequency of an oscillator could be used as a vibrato effect.

LFOs are otherwise identical to oscillators. They can vibrate in different shapes, and you can control their frequency. Typically, because LFOs are not audio generators, the frequency control will be labeled "Rate," not "Pitch."

To learn the many ways LFOs can be used, see Chapter 6.

ENVELOPE GENERATORS

An envelope generator controls how a sound changes over time. While an LFO is a modulation source that loops over and over, an envelope generator provides a one-shot voltage signal that describes how a parameter changes from start to finish.

Often, an envelope generator is used to control a VCA. Imagine the sound of a bowed violin getting louder over the course of a few seconds versus the sound of a plucked string on the same violin. The difference between the two sounds is largely that of the envelope—the former starts quiet and gets progressively louder, while the latter starts and stops quite abruptly.

Many commercial synthesizers have an envelope with four potentiometers labeled ADSR, which stands for "attack, decay, sustain, and release." These four parameters let us describe the shape of how we want that note to change over time. While these often apply to the VCA, an envelope's shape can be used to modulate any other parameter. You could use an envelope generator on a filter, for example, to cause the sound's harmonics to deaden or lighten as the note progresses. You could also use it on an oscillator's pitch control, causing a note to bend up or down.

To make a simple envelope-like effect that controls a VCA, see Chapter 10.

SAMPLE-AND-HOLD GENERATORS

Sample-and-hold circuits are confusingly named. They input the voltage of a continuously varying analog signal, such as an oscillator, or noise generator. When the circuit receives a signal, it will sample (i.e., take an instantaneous snapshot of the voltage level) and then hold that voltage until it is retriggered. These are a great way to make R2-D2-style *bleep-bloop*s.

To build a sample-and-hold generator, see Chapter 13.

CONTROL SYSTEMS

Many synthesizers are designed to be played in real-time, which makes control interfaces very important. A control system is simply a way for a human to easily input information into the synth—for instance, with a piano keyboard. There are numerous types of control systems, and we encourage you to invent your own, but here are a few:

KEYBOARDS

A keyboard is not a synthesizer. A keyboard is a user interface that can be used to control a synthesizer.

When you press a key on a keyboard-controlled synth, you are typically doing two things: First, pressing the key typically turns up the amplifier, which makes the sound audible—press a key, and you hear a sound. Second, the key you press sends a specific voltage level to an oscillator. This control voltage is what tunes the oscillator. Touch a higher key, and you'll get a higher voltage level and a higher pitch.

Building a keyboard isn't hard—take a series of tact switches (momentary-on push buttons) and put them in your breadboard. When each one is toggled, have it complete an oscillator circuit via a potentiometer. However, because the focus of this book isn't on building traditional instruments, we have not included a formal write-up in this book. But don't let that stop you—try it out, and let us know how it went!

SEQUENCERS

A sequencer reads back a series of voltages in a sequence. Often, you'll see them as melodic controllers that quickly cycle through a series of potentiometers that retune an oscillator. If you hear a synth play a repeated line really quickly, chances are it's not a human playing a keyboard but a sequencer happily cycling along.

To build a sequencer, see Chapter 11.

RIBBON CONTROLLERS

A ribbon is a long, skinny controller that can be played with a single finger. The ribbon itself acts like a potentiometer, and pressing it at a certain place will cause the circuit's resistance to immediately change to a new value.

You can build a ribbon controller with a piece of VHS tape, with the two ends of the tape acting as the "cheeks" of a potentiometer, and a third gator clip (used to touch the ribbon) that acts as the "nose," or wiper, of the potentiometer. This setup works quite well with the VCO described in the Phase-Locked Loops chapter. Set up your ribbon as a voltage divider, and you're good to go.

Ribbon controllers, while less-common nowadays, were quite common in early electronic music: classic instruments such as the ondes martenot and trautonium both operate on this very principle!

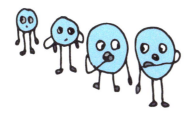

APPENDIX II

INTEGRATED CIRCUIT INFORMATION

T his section details the basic operation of the most essential integrated circuits found in this book, their potential musical uses, and other vital information about how to get them working.

CD40106 HEX SCHMITT-TRIGGER INVERTER

What it is:
A six-pack of inverters all available on a single chip.

How can we use it?
- You can use an inverter to change a high voltage into no voltage (or vice versa), thus flipping a square wave upside down.
- You can also use an inverter to make a simple square, triangle, or sawtooth oscillator.

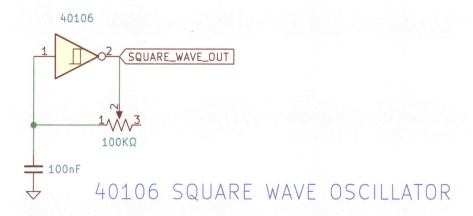

40106 SQUARE WAVE OSCILLATOR

Basic operation:
- Send VCC to pin 14, and ground to pin 7. Boom—you've set up six inverters. Any voltage low going to pin 1 will be inverted to voltage high and come out of pin 2. The same is true for pins 3 and 4, pins 5 and 6, pins 8 and 9, pins 10 and 11, and pins 12 and 13.
- To set up an inverter as a square-wave oscillator, attach a capacitor to the inverter's input. Attach a resistor (or potentiometer, for playability) between the input and the output. Amplify the output to hear your square wave.
- To set up an inverter as a triangle-wave oscillator, build a square-wave oscillator and amplify the *input*. Instead of a square wave turning harshly on and off, we'll hear the capacitor slowly charge and discharge—a much more triangular shape!

- To set up an inverter as a sawtooth oscillator, build a square-wave oscillator and connect a diode between the input and the output. Because a diode allows current to flow in only one direction, the shape of our wave will resemble a ramp more so than a flat-topped square.

Some inspiration on what can be done with this chip:
- Build six oscillators with photoresistors instead of potentiometers. While not exactly easy to control, this is a shockingly fun thing to play around with.

Notes:
- You'll notice your square wave sounds a lot louder than the triangle or sawtooth waves that come out of this chip. If you'd like to use these waves together, we recommend putting different value resistors between the waves and the mixer so you can add different volumes of waves to taste.

CD4093 QuAD SCHMITT-TRIGGER NAND GATE

What it is:
A four-pack of NAND gates on a single chip.

How can we use it?
- Just like the 40106, a NAND gate can be hooked up as an inverter simply by attaching one of the inputs to VCC.
- Not only can you use a NAND gate as a simple oscillator, its additional input pin also gifts you an easy modulation source that can turn your wave on and off. By attaching one square wave to the input of another, you can get an oscillator that beeps like a smoke detector.

Basic operation:
- Send VCC to pin 14, and ground to pin 7. Pins 1, 2, and 3 comprise one NAND gate with two inputs and one output, respectively. Pins 6, 5, and 4 comprise a second NAND gate, with two inputs and one output, respectively. Pins 8, 9, and 10 comprise the third NAND gate, with two inputs and one output, respectively. Pins 13, 12, and 11 comprise the final NAND gate, with two inputs and one output, respectively.
- To use a NAND gate as an inverter, attach one of the input pins to VCC. Now, the output pin will always exhibit the inverse of the input pin.

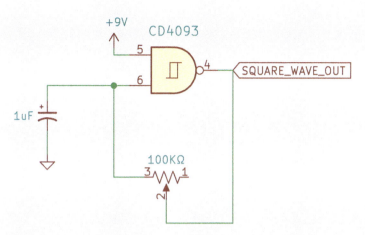

4093 SQUARE WAVE OSCILLATOR

- To set up a NAND gate as an oscillator, attach one input to VCC, a capacitor between the second input and ground, and a resistor (or potentiometer, for playability) between the input with the capacitor and the output. Amplify the output to hear your square wave. Notice how when that VCC input is switched to ground, the oscillator stops.
- You can also use this oscillator as a triangle wave (listen to the capacitor input) or a sawtooth wave (connect a diode between the input and output).
- To chain one square-wave oscillator into another, send the output of a square wave to the input pin of another NAND gate oscillator. By sending a signal of alternating voltages to the other oscillator, we can make it turn on and off. See Chapter 10 on chaining oscillators for countless more things you can do with this method.

CD4017 DECADE COUNTER
What it is:

The little chip that counts! The 4017 has an input for a square wave and has ten pins that operate like fingers. Whenever the square wave ticks forward by one period, the 4017 sends current out of the next "finger." The IC also comes with a "reset pin," which starts the count over whenever it receives a VCC-level signal. By attaching any of the counter pins to the reset pin, we can make the 4017 count to any number less than ten.

How can we use it?

The most basic circuit utilizing a 4017 is likely an LED-chaser circuit. With ten LEDs attached to the counting pins and a low-frequency square wave going into the input, we can make our ten LEDs light up one at a time.

A more musical usage of this chip is to use it to divide frequencies—see more in Chapter 9 on harmonization.

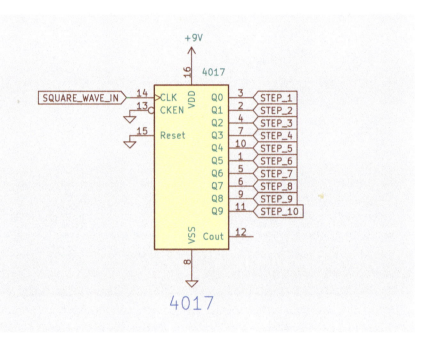

Basic operation:

- Send VCC to pin 16 and ground to pin 8. We will also need to connect pins 13 and 15 to ground. Send a square wave to pin 14.
- Attach ten resistors and LEDs to the counter pins, and watch as they light up one at a time.
- To make the 4017 count to a number less than 10, attach pin 15 to the counter pin that corresponds to the "final number" you're counting to.

CD4040 BINARY COUNTER AND DIVIDER
What it is:

A CD4040 takes a square wave input and slows it down by a factor of two. Then it takes that slowed-down wave and slows *that* down by a factor of two. And again, and again.

How can we use it?

- By putting an audio-rate square wave into the 4040's input, our chip automatically generates a rich array of sub-octaves.
- If we provide an LFO-rate square wave, our chip will generate nested binary rhythms.
- We can use three consecutive divisions to count in binary—our 4051 multiplexer chip was designed specifically for this integration. This binary count is digital information that our ICs understand how to work with.

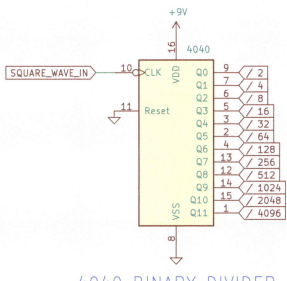

4040 BINARY DIVIDER

Basic operation:

Send VCC to pin 16 and ground to pin 8. We will also need to connect pin 11 to ground. Send a square wave to pin 10. Use your amplifier to listen to the outputs, and notice they're totally in-tune octaves.

CD4051 EIGHT-CHANNEL MULTIPLEXER
What it is:

A very cool and often-misunderstood chip. The simplest way to grasp the CD4051 is to think of an eight-way switch. We can direct one stream of current to one of eight possible outputs. Inversely, we can also take eight streams of current and switch between them, connecting certain signals to one common output channel.

How can we use it?

We can use the 4051 to automatically switch between multiple audio signals or control voltages. This is particularly useful for sequencers, where we want to automatically switch between different voltage levels.

The two examples I like to give involving the 4051 are a little ridiculous. We could attach eight speakers to the outputs of the 4051, send a piece of music into the input, and tell the 4051 to very quickly jump between the speakers. Now, we have a circuit where one piece of music jumps around to various corners of your room.

The second example is the same circuit in reverse. Instead of eight outputs, what if we had eight inputs? We could send eight songs to the 4051, attach one speaker to the output, and then have the 4051 rapidly switch between songs. Now, we have a circuit that channel-surfs between various songs at breakneck speed.

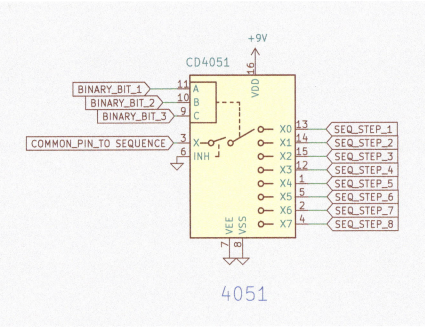

Basic operation:

- Send VCC to pin 16 and ground to pin 8. We will also need to connect pins 6 and 7 to ground.
- Pins 11, 10, and 9, respectively, are three digits of a binary number. This binary number is the "address" that tells the IC which connection to make internally. Attaching one of these pins to VCC will code as a 1, and attaching it to ground will code as a 0. If we want to attach pin 3 (our common pin) to

pin 15 (labeled X2), we will have to use pins 11, 10, and 9 to communicate the number 2 in binary. The number 2, which is 010 in three-digit binary, can be programmed by hooking up pin 11 to ground, pin 10 to VCC, and pin 9 to ground (forming a 0-1-0). As soon as our address pins are hooked up, the 4051 will do its thing and connect pin 3 to pin 15.

- We can make the 4051 count in numerical order by sending three consecutive 4040 outputs to 4051 pins 11, 10, and 9. These cascading square waves actually produce a binary number sequence when read out. For more information, see Chapter 11 on sequencers.

- While a 4040 produces a clean counting system, it should be noted that *any binary signal* will result in the 4051 making an internal connection. For instance, if you made three square-wave oscillators and hooked each one up to an address pin, every time a square wave oscillated, your 4051 would switch its connection. This is a great way to make a really unpredictable sequencer.

CD4046 PHASE-LOCKED LOOP
What it is:

A chip with so many uses, we dedicated a whole chapter to it! The 4046 has a built-in VCO (voltage-controlled oscillator), which it can synchronize with an external square wave. If we incorporate a frequency divider (such as the 4017), we can generate multiples of the input frequency.

How can we use it?

- A VCO, while seemingly similar to our 40106 and 4093 oscillators, is a little different in terms of execution. Instead of the resistance between two places driving the frequency, the pitch is determined by the electrical potential of a single pin. This means we can play with our oscillator with a voltage divider, or even supply an external triangle or sawtooth wave to modulate our oscillator's pitch.

- Another advantage of a VCO is that we can add a portamento (glide) when we switch between notes—all we have to do is add a large capacitor between the VCO input and ground. A larger capacitor will result in a longer glide time.

- The frequency-tracking capabilities of the 4046 also let us *multiply* frequencies—see Chapter 13 on phase-locked loops for more information.

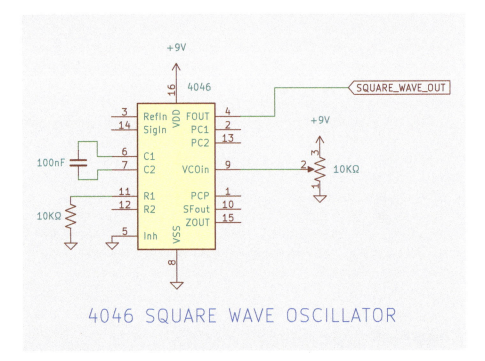

4046 SQUARE WAVE OSCILLATOR

Basic operation:

- Send VCC to pin 16 and ground to pin 8. We will also need to connect pin 5 to ground.
- Add a resistor between pin 11 and ground and a capacitor between pins 6 and 7. The value of these two components will determine the range of your oscillator.
- Create a voltage divider with a potentiometer by attaching the cheeks to VCC and ground. Send the nose to pin 9, which is our VCO's input.

LM386 POWER AMPLIFIER

What it is:

It's a classic, famously cheap, easily accessible IC for making your own power amplifier. It's kinda loud, it's kinda functional, and you'll have kinda fun.

How can we use it?

A power amplifier is a circuit that can take a really low-current signal (such as a 4093 square wave oscillator) and beef it up into a high-enough current level to easily move a speaker.

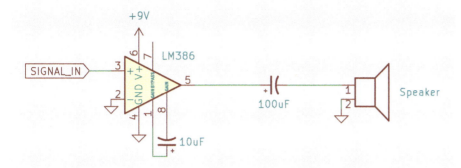

POWER AMPLIFIER

Basic operation:

- Connect pin 6 to VCC. Connect pins 2 and 4 to ground.
- Attach a 10 µf capacitor between pin 1 and pin 8 and the positive leg of a 100 µf capacitor to pin 5. Leave the other end in an empty breadboard row.
- Connect the signal you want to amplify to pin 3. Attach one tab of a speaker to the spare leg of the 100 µf capacitor and the other to ground.
- A volume knob can be added with a logarithmic potentiometer. It can be placed either between the signal source and pin 3 or between the output capacitor and the speaker.

TL074 OP AMP
What it is:

An easy operational amplifier, great for use in audio mixers and active filters.

How can we use it?

Send a weak signal in, get a stronger signal out. This will make for a much more stable mix of signals than simple passive potentiometers. Active filters differ from passive filters in that they have an amplification stage.

It should be noted that although the TL074 and the LM386 are both ICs designed for amplification, they have different purposes. *Your TL074 will amplify your inputs to make a clean mix, but it does not itself have the current necessary to drive a speaker.* A power amplifier, like the LM386, is designed to output high-enough currents to drive a speaker.

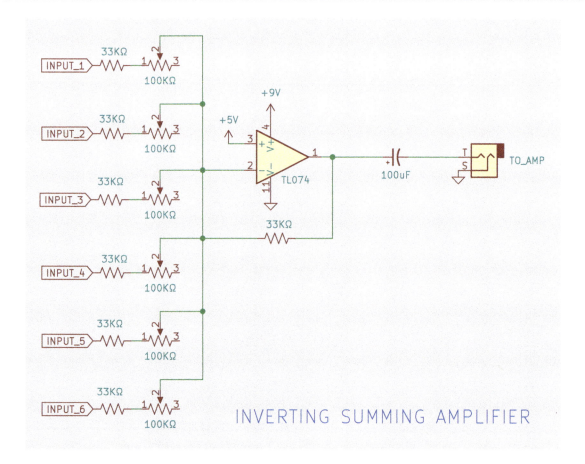

INVERTING SUMMING AMPLIFIER

Basic operation:

- For our basic power connections, pin 4 attaches to VCC and pin 11 attaches to ground.
- To build a mixer, send an arbitrary number of channels through an arbitrary number of potentiometers. Send the combined output of the potentiometers to both pin 8 and pin 9.
- To center our amplified output around a +5 V potential, use a voltage regulator to send 5 V to pin 10, a noninverting input.
- Connect pin 8 and pin 9 to each other through a 33 k ohm resistor.
- Add a 100 µf capacitor to the output and your power amplifier.

CD4053 3-CHANNEL ANALOG MULTIPLEXER

What it is:

Three train-track-style switches that can be independently toggled. Pathway 1 can go to A or B, pathway 2 can go to C or D, and pathway 3 can go to E or F.

How can we use it?

It's an easy way to get one signal to jump between two places.

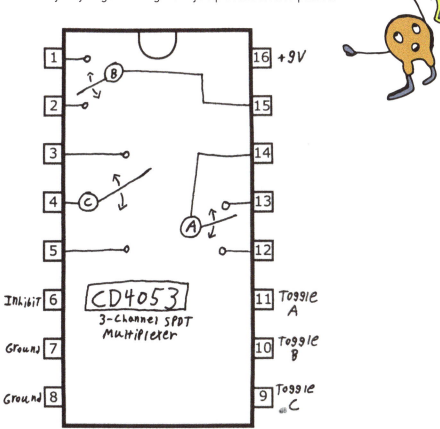

Basic operation:

- Connect pin 16 to VCC and pins 6, 7, and 8 to ground.
- Pin 14 can connect to pin 13 or pin 12 based on the toggle state of pin 11.
- Pin 15 can connect to pin 1 or pin 2 based on the toggle state of pin 10.
- Pin 4 can connect to pin 3 or pin 5 based on the toggle state of pin 9.

CD4066 QUAD BILATERAL SWITCH

What it is:

A four-pack of simple electronic toggles. A square wave will repeatedly close and open a circuit for you!

How can we use it?

If you want a circuit to repeatedly press a button for you, this is your chip.

Basic operation:

- Pin 14 goes to VCC and pin 7 goes to ground.
- Pin 13, when toggled, will connect pin 1 to pin 2.
- Pin 5, when toggled, will connect pin 3 to pin 4.
- Pin 6, when toggled, will connect pin 8 to pin 9.
- Pin 12, when toggled, will connect pin 10 to pin 11.

CD4070 QUAD XOR GATE

What it is:

A quartet of XOR gates—logic gates that output current *when only one of the inputs receives current*.

How can we use it?

XOR gates are ideal for making inharmonic sounds, which is precisely how traditional electronic drum machines emulate a cymbal.

Basic operation:

- Pin 14 connects to VCC and pin 7 connects to ground.
- Pins 1 and 2 are the inputs for the first XOR gate, which outputs to pin 3.
- Pins 5 and 6 are the inputs for the first XOR gate, which outputs to pin 4.
- Pins 8 and 9 are the inputs for the first XOR gate, which outputs to pin 10.
- Pins 12 and 13 are the inputs for the first XOR gate, which outputs to pin 11.

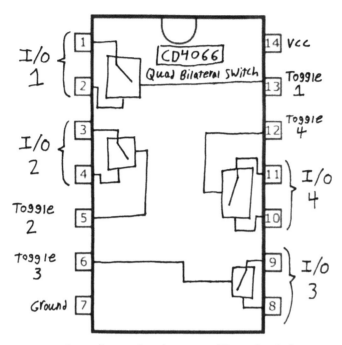

Pinout diagram for the CD4066 bilateral switch.

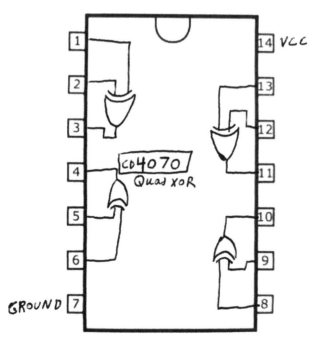

Pinout diagram for the CD4070 XOR gate.

APPENDIX III

PART SOURCING

There was once a time, just a few decades ago, when most North American cities had a place where you could buy individual electronic components. These days, the economic model just doesn't allow for it. Few people are building electronics at home, the cost of the items is too cheap to justify, and the small numbers of parts that hobbyists purchase generally isn't enough to keep the lights on.

While many parts of the world, notably Latin America, still seem to have thriving markets for electronic components, there's a fair chance you might not have that resource. Fortunately, the good old internet has plenty of suppliers for parts that are easy to navigate, and cheap.

COMPONENT SUPPLIERS

The big ones:

- Digikey (Digikey.com) and Mouser (Mouser.com) are stocking distributors, which has nothing to do with footwear. Both companies sell tens of thousands of components and have handy websites with complex search features to ensure that you're getting the part you need. If neither of these websites has the part you're looking for, there's a chance it doesn't exist.
- Jameco (Jameco.com) is a California-based retailer of parts that is well known in the hobbyist market. It sells plenty of prepacked kits with all the parts you need inside.
- Adafruit (Adafruit.com) is a boutique hobbyist shop that supplies all kinds of exotic sensors, speakers, and other doohickies. It stocks a large supply of cool-looking switches, LEDs, and supplies for enclosures.

Smaller ones:

- Tayda (Taydaelectronics.com) is a small company that sells the B stock from larger companies such as Digikey and Mouser. However, it positions itself specifically for the hobbyist demographic. As of this writing, it ships small orders from Colorado and larger orders from Thailand.
- AllElectronics (Allelectronics.com) is an LA-based new-surplus supplier, meaning it sells overstock from other companies.

Some tips:

- Buy in bulk. Getting one resistor might cost you only a nickel, but getting a hundred might cost only a dollar. Not only will you save a little money, but it's also always helpful to have extra components around.

- Make sure you're buying through-hole components and *not SMD components*. Through-hole components are designed for human-size hands. In contrast, SMD (or surface-mounted) components are incredibly small and are designed for machines to put them together. It's very easy to accidentally get an SMD-format capacitor or resistor, which won't be of much use to you. Avoid making this mistake.
- Similarly, if you're buying any integrated circuits (such as 555 timers), make sure the product says *DIP* (dual-inline package). These chips are the right size to fit in breadboards.
- If you want to keep things simple, take a look at the Maker Shed (MakerShed.com)—we're working with our friends at Make: to provide kits for our projects. ♪

OHM'S LAW

Ohm's law is the bread and butter of electronics, which lets us describe the relationship between voltage, current, and resistance.:

$V = I \times R$

- V = voltage (in volts)
- I = current (in amps)
- R = resistance (in ohms)

For instance, let's say we want to calculate the value of a resistor we need to light up a 0.02 amp LED with a 9 V power supply. We can rearrange Ohm's law to form R = (V/I) and evaluate the function as R = (9/0.02) = 450 Ω . A 450 Ω resistor will do the trick!

CALCULATING POWER

The power of a circuit (in watts) is represented by P, and it describes the amount of energy per second that the circuit is using:

$P = V \times I$

A circuit that consumes 9 V at 2 amps would produce 9 x 2 watts of power, or 18 watts.

RESISTORS IN SERIES

If we put a bunch of resistors back-to-back, we can make a bigger resistor—just add up the resistances! For instance, if we have two 100 Ω resistors, we can connect them to form one 200 Ω resistor.

If our individual resistors are R1, R2, and R3, and our sum is Rtotal, the relationship looks like this:

$Rtotal = R1 + R2 + R3$

RESISTORS IN PARALLEL

The calculation gets a little more complex when the resistors run parallel to each other. Instead of one resistor emptying into another, a parallel configuration allows for electrons to travel down one of many paths.

If our individual resistors are R1, R2, and R3, and our sum is Rtotal, the relationship looks like this (I hope you like fractions):

Rtotal = 1 / ((1/R1)+(1/R2)+(1/R3))

CAPACITORS IN SERIES

Somewhat perplexingly, the process for calculating capacitance with many capacitors is the opposite of how you would do it with resistors. If we were to put a number of capacitors back-to-back, we could calculate their series capacitance as follows.

If our individual capacitors are C1, C2 and C3, and our sum is Ctotal, the relationship looks like this:

CTotal = 1/((1/C1)+(1/C2)+(1/C3))

CAPACITORS IN PARALLEL

If our individual capacitors are C1, C2 and C3, and our sum is Ctotal, the relationship looks like this:

Ctotal = C1 + C2+ C3

RC TIME CONSTANTS

Some circuits have a *time constant* that will affect some frequencies but not others. In our chapter on filters, we spent considerable time playing around with circuits built from resistors and capacitors, called RC circuits. Capacitors take a short amount of time to fill and more time to drain through a resistor. By putting in larger resistors of capacitors, we can extend the discharge time.

The RC time constant formula is how to calculate the number of seconds it will take for the circuit to respond:

T = R x C

- T = time (in seconds)
- R = resistance (in ohms)
- C = capacitance (in farads)

For example, let's evaluate the time constant for a circuit with a 1 kΩ resistor and a 0.1 μf capacitor. A 1 kΩ resistor is 1,000 Ω, and a 0.1 μf capacitor is equal to .0000001 farads. Multiplying those two, we get 0.001 seconds (or 1 millisecond).

VOLTAGE DIVISION

You can use a potentiometer (or two fixed-value resistors) to create a voltage divider from a higher input voltage:

$$Vout = Vin \times ((R2)/(R1+R2))$$

- Vout = divided voltage output (in volts)
- Vin = whole voltage input
- R1 = resistor 1 (in ohms)
- R2 = resistor 2 (in ohms)

In the case of a potentiometer, R1 and R2 will always sum to the total value of your potentiometer.

VOLTAGE GAIN

To calculate the gain of an inverting op amp (like the TL074), the equation is as follows:

$$Gain = Vout/Vin = -RF/RI$$

- Gain = proportion of change (as a proportional number)
- Vout = voltage at output (in volts)
- Vin = voltage at input (in volts)
- RI = resistance of input resistor
- RF = resistance of feedback resistor

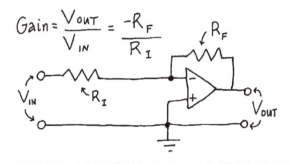

These are ten pieces of electronic music that showed me what electronic music *could* be. This doesn't mean that they all sound good. Not all of the following pieces are even "synthesizer based." But that's not really the point—as we've waxed poetic about this before, music technology is a cultural practice more than a technical one. Each of these works taught me something new about what electronic music could communicate, and hopefully, they'll be inspirations to you as well.

Halim El-Dabh, "The Expression of Za'ar" (composition, 1944)

While working for a Cairo radio station, composer Halim El-Dabh made ample use of some top-of-the-line audio equipment while his boss wasn't watching. After sneaking off with a wire recorder (a proto tape recorder that encoded signals onto a magnetizable wire), El-Dabh was able to record a *zār*, a traditional public spirit-possession ritual.

Back in the studio, the sounds of the ritual were processed so as to creatively allude to something supernatural. The piece makes use of plate reverbs (like the one we built in Chapter 4 on amplifiers), and VCAs (like the one in Chapter 10 on modulation) to stunning effect. While the piece isn't technically very complex, it represented a big conceptual leap in how recording technology could be used to compose. In this case, El-Dabh recorded something and transformed it into something *it wasn't*.

Pierre Henry, *Variations pour une porte et un soupir* (album, 1963)

A piece of music made almost exclusively from two sounds—a person sighing and a door creaking. That's the concept. Pierre Henry (who supposedly spent two hours a day practicing the "creak" to get it just right) highlights the similarities between the two sounds as he plays them in series, then on top of each other, then removes one entirely, then brings it back in a cacophonous chorus. By the end, you'll likely be confused as to which sound you're listening to. The work perplexed a lot of the French musical establishment, who at the time were still getting used to the notion of music *without* traditional instruments—even today, this piece raises a few eyebrows. However, the basic recipe used by Henry is evergreen and worth experimenting with. Try recording a few sounds of your own that sound similar. How would you arrange them to draw attention to their similarities?

Alvin Lucier, *I am sitting in a room* (performance, 1969)

This simple performance begins with a reading of a short paragraph of text:

I am sitting in a room different from the one you are in now. I am recording the sound of my speaking voice and I am going to play it back into the room again

and again until the resonant frequencies of the room reinforce themselves so that any semblance of my speech, with perhaps the exception of rhythm, is destroyed. What you will hear, then, are the natural resonant frequencies of the room articulated by speech. I regard this activity not so much as a demonstration of a physical fact but more as a way to smooth out any irregularities my speech might have.

Didactic as it may seem, it takes Lucier only a few readings and rerecordings for the sound to turn into a pulsing hum that barely resembles human speech. The experience, while a little slow, is pretty incredible to behold—you can try it yourself at home, in any room! A simple electronic process of recording, amplification, playback, and rerecording is used with great effect to turn something human sounding into something very inhuman sounding.

Bruce Haack, *The Electric Lucifer* (album, 1970)

I first learned of Bruce Haack from an appearance on *Mr. Rogers' Neighborhood*, of all places, where he shows off his homemade musical instruments and invites kids to dance. By most accounts, Haack really only *kind of* understood how his synthesizers worked (he was, after all, more interested in the music than the engineering), yet his real skill was figuring out the narrative context that an instrument would fit in. *The Electric Lucifer* is the pinnacle of Haack's musical output, which uses somewhat unpredictable synthesizers to illustrate a metaphysical battle between angels and demons. Haack processes his voice to make each character sound different, and takes you on a literal journey to hell through a menagerie of homemade electronics. All of Haack's output is worth a listen, but this is the quintessential place I'd encourage you to start.

Lene Lovich, *Stateless* (album, 1978)

Born in Michigan, Lene Lovich had her big break in England when she was hired to dub screams for horror movies. Her deliciously idiosyncratic voice came full circle in the popular imagination of electronic music—she sings alongside, and often emulates, the synthesizers fast becoming au courant at the time. Lovich chirps, wails, and whistles through a selection of exquisitely crafted pop songs unlike anything you've ever heard. Why she never became a household name is beyond me.

Laurie Anderson, "Late Show," *Home of the Brave* (film, 1986)

No book of electronic music would be complete without a passing mention of composer/inventor Laurie Anderson. For a great overview of her abilities, I humbly suggest the concert film *Home of the Brave*, which covers quite a bit of her artistic territory. "Late Show" features a particularly enigmatic invention, a violin interface with a bow made of magnetic audio tape. As she bows the instrument, she drags the tape past a playhead embedded in the violin body and *manually* plays a sample. In a poetic reversal that would have made Pierre Henry's academy squirm, Anderson ties together the physical interface of a traditional musical instrument with the sample playback of a tape player. This performance prompts the question "If a tape player isn't a musical instrument, does a violin interface negate that assertion?"

Carl Stone, "Shing Kee" (composition, 1986)

Carl Stone makes sound out of other sounds, often using a very logical process. In the case of "Shing Kee," we hear a small fragment of a piano-and-voice piece that loops over and over. With every loop, however, the fragment is extended by a fraction of a second, and the loop slowly, and with hypnotic effect, changes to encompass more and more musical material. The second half involves a second looping musical fragment that gets slowly time-stretched with every reintroduction. The piece draws your attention to the artifice of recording in an incredibly simple but beautiful way. Most of Stone's output involves some sort of algorithmic process, and figuring out the process is earnestly part of the fun.

Pamela Z, *Echolocation* (album, 1987)

Since she started working in the mid 1980s, Pamela Z has built a career around her voice and how it's represented. Her performances mix spoken word with electronic manipulation, letting the listener draw parallels between *what* she says and *how* she says it. *Echolocation* is Z's first album, a particularly lovely set of songs that pine for, in the artist's words, "sounds yet unfelt."

Also highly recommended are Pamela's shows with sensor-laden mechanical co-performers. You can see how her slight gestures control the electronic effects with stunning clarity.

Negativland, *Helter Stupid* (album, 1989)

There's no sufficient way to describe the experimental band Negativland, out of El Cerrito, California, but this album might give you an idea. As part of a convoluted prank, band members sent (very fake) press releases to San Francisco Bay Area TV stations claiming that their previous album was so controversial that it led to a violent family homicide in Ohio. Without fact-checking, several news outlets actually ran this story, giving Negativland more attention than they had ever realistically assumed they would achieve. To sell the illusion, the band members kept totally silent about it and had a friend pose as an FBI agent to do on-air interviews. What was the purpose of the prank? All the news coverage was recorded, remixed, and turned into a *new* album. *Helter Stupid* is an obnoxious and strange poetic collage about the state of news media in 1989 that plays just as well today. Is it brilliant? Is it shockingly immoral? That's up for you to decide.

Juana Molina, *Halo* (album, 2017)

As Argentine singer Juana Molina's output expands, she seems to only grow fonder of synthesizers. This album, filled with hauntingly beautiful and lyrically surreal imagery, supplies pop structures with an undeniably sinister undertone. The album marries Molina's breathy, folky voice with looping synths, off-kilter rhythms, trancelike pulsating pads, odd time-signature changes, and onomatopoetic mutterings. It's a breathtakingly refreshing synthesizer album that doesn't really sound much like a synthesizer album.

APPENDIX VI

SEVENTY GREAT SYNTHESIZER ALBUMS

- Delia Derbyshire, *Inventions for Radio: Dreams* (United Kingdom, 1964)
- Raymond Scott, *Soothing Sounds for Baby* (United States, 1964)
- Perrey and Kingsley, *The In Sound from Way Out!* (France, 1966)
- Morton Subotnick, *Silver Apples of the Moon* (United States, 1967)
- The Monkees, *Pisces, Aquarius, Capricorn & Jones Ltd.* (United States, 1967)
- Silver Apples, *Contact* (United States, 1968)
- Terry Riley, *A Rainbow in Curved Air* (United States, 1969)
- Herbie Hancock, *Head Hunters* (United States, 1973)
- Can, *Future Days* (Germany, 1973)
- Eno, *Here Come the Warm Jets* (United Kingdom, 1974)
- Tangerine Dream, *Phaedra* (Germany, 1974)
- Barış Manço, *2023* (Turkey, 1975)
- Biddu & the Orchestra, *Eastern Man* (United Kingdom/India, 1977)
- William Onyeabor, *Who Is William Onyeabor?* (Nigeria, 1977–1985)
- Giorgio Moroder, *From Here to Eternity* (Italy, 1977)
- Yellow Magic Orchestra, *Technodelic* (Japan, 1978)
- Nina Hagen Band, *Unbehagen* (Germany, 1979)
- Sparks, *No. 1 in Heaven* (United States, 1979)
- Bill Nelson's Red Noise, *Sound on Sound* (United Kingdom, 1979)
- Telex, *Looking for Saint Tropez* (Belgium, 1979)
- Curtis Mayfield, *Something to Believe In* (United States, 1980)
- Laurie Spiegel, *The Expanding Universe* (United States, 1980)
- Kraftwerk, *Computer World* (Germany, 1981)
- Oingo Boingo, *Only a Lad* (United States, 1981)
- Alice Coltrane, *Turiya Sings* (United States, 1982)
- Haruomi Hosono, *Philharmony* (Japan, 1982)
- Thomas Dolby, *The Golden Age of Wireless* (United Kingdom, 1982)
- George Clinton, *Computer Games* (United States, 1982)
- Suzanne Ciani, *Seven Waves* (United States, 1982)
- Laurie Anderson, *Big Science* (United States, 1982)
- Aviador Dro, *Alas Sobre El Mundo* (Spain, 1982)
- Charanjit Singh, *Synthesizing: Ten Ragas to a Disco Beat* (India, 1982)
- Brian Eno, Daniel Lanois, and Roger Eno, *Apollo: Atmospheres and Soundtracks* (Canada, 1982)
- Francis Bebey, *Psychedelic Sanza* (Cameroon, 1982–1984)
- Steve Roach, *Structures from Silence* (United States, 1984)
- Robert Gorl, *Night Full of Tension* (Poland, 1984)

- Severed Heads, *Since the Accident* (Australia, 1984)
- Grace Jones, *Slave to the Rhythm* (Jamaica, 1985)
- Hiroshi Yoshimura, *Green* (Japan, 1986)
- Frank Zappa, *Jazz From Hell* (United States, 1986)
- The KLF, *The White Room* (United Kingdom, 1991)
- Aphex Twin, *Selected Ambient Works 85–92* (United Kingdom, 1992)
- Stereo Nova, *Asyrmatos Kosmos* (Greece, 1993)
- Portishead, *Dummy* (United Kingdom, 1994)
- The Future Sound of London, *Lifeforms* (United Kingdom, 1994)
- Massive Attack, *Protection* (United Kingdom, 1994)
- The Chemical Brothers, *Exit Planet Dust* (United Kingdom, 1995)
- Stereolab, *Dots and Loops* (France, 1997)
- Björk, *Homogenic* (Iceland, 1997)
- Aterciopelados, *Caribe Atómico* (Colombia, 1998)
- Susumu Yokota, *Sakura* (Japan, 1999)
- Pepe Deluxe, *Super Sound* (Finland, 1999)
- Boards of Canada, *Music Has the Right to Children* (Scotland, 1998)
- Röyksopp, *Melody A.M.* (Norway, 2001)
- Broadcast, *Haha Sound* (United Kingdom, 2003)
- Air, *Talkie Walkie* (France, 2004)
- Nortec Collective, *Tijuana Sessions, Vol. 3* (Mexico, 2005)
- Juana Molina, *Un Día* (Argentina, 2008)
- Dewanatron, *Semi Automatic* (United States, 2009)
- Black Moth Super Rainbow, *Eating Us* (United States, 2009)
- The Bird and the Bee, *Ray Guns Are Not Only Just the Future* (United States, 2009)
- Bottlesmoker, *Hypnagogic* (Indonesia, 2013)
- Omar Souleyman, *Wenu Wenu* (Syria, 2013)
- Oneohtrix Point Never, *R Plus Seven* (United States, 2013)
- Janelle Monáe, *The Electric Lady* (United States, 2013)
- Anderson .Paak, *Malibu* (United States, 2016)
- Kaitlyn Aurelia Smith, *The Kid* (United States, 2017)
- Thundercat, *Drunk* (United States, 2017)
- Vansire, *Angel Youth* (United States, 2018)
- Kate NV, *Room for the Moon* (Russia, 2020)

APPENDIX VII

GLOSSARY!

AC (alternating current):
Electric current that reverses direction periodically, most familiar as the current found in your household electrical outlet. The frequency is typically 60 times a second in North America and 50 times a second elsewhere in the world.

Active component:
A circuit or component that requires an external power source to operate, such as an IC, transistor, or LED. Contrast with "passive component."

Additive synthesis:
A method of sound synthesis in which complex sounds are generated by combining multiple sine waves at different frequencies and amplitudes.

ADSR:
An abbreviation of Attack, Decay, Sustain, and Release, four stages seen in many commercial envelope generators. The ADSR envelope shapes the characteristics of a sound over time, controlling parameters such as volume, timbre, and amplitude.

Alligator clip:
A swampy-themed, spring-loaded metal clip used to make temporary electrical connections.

Alloy:
A mixture of two or more metals, or a metal and another element, to enhance properties like electrical conductivity or tensile strength.

Ampere (A):
The unit of electric current, representing the flow of electric charge.

Ampere's law:
A fundamental law of electromagnetism that states that the magnetic field around a closed loop of wire is directly proportional to the electric current passing through the wire.

Amplifier:
A device that increases the power of a signal.

Amplitude:
The maximum extent of a vibration or oscillation, measured from the position of equilibrium. It can be measured in volts (if an audio signal) or in pressure levels (if a sound).

Amplitude modulation:
The act of using the shape of one wave to change the amplitude of another wave.

AND gate:
A logic gate that outputs true (i.e., spits out current) only when
all of its inputs are true (i.e., receiving current).

Attenuation:
The act of reducing the gain of a signal.

Atom:
The basic unit of a chemical element, composed of a nucleus containing
protons and neutrons and surrounded by electrons.

Audio:
Sound within the range of human hearing, typically between 20 Hz and 20 kHz if
you have textbook ears and 60 Hz through 15 kHz if you are a normal human.

Balanced audio:
A method of interconnecting audio equipment using cables that carry two signals 180 degrees
out of phase with one another, along with a ground wire. Opposite polarities are used for the
signals, which allows the receiving equipment to cancel out any noise picked up along the cable by
inverting the phase of one signal. This method is commonly used in professional audio setups to
maintain high-quality sound over long distances.

Beat:
The fundamental counting unit in music that establishes a tempo. A beat may be audible
or implied.

Binary:
A numbering system based on base 2 in which the only digits are 0 and 1.

Boolean:
A type of algebraic logic that deals with variables and operations such as AND, OR, and NOT.

Breadboard:
A solderless device used to prototype electronic circuits, consisting of a
perforated board with spring clips for connecting components.

Capacitor:
An electronic component that stores electrical energy in an electric field.

Channel (audio):
An independent path allowing audio signals to flow, typically referring to a single audio input or output.

Charge:
A force that comes in two flavors—positive and negative. Charges of the same flavor repel each other, while opposite flavors attract. Although we don't really understand *what* charge is, we know it is a fundamental property of matter responsible for electrical phenomena.

Circuit:
A closed loop through which an electric current can flow. Just like a circle, a circuit must be connected to itself.

Circuit board:
See "printed circuit board (PCB)."

CMOS (complementary metal-oxide semiconductor):
A technology used in integrated circuits to construct digital logic gates and other components. Almost all the ICs we use in this book are CMOS, as is the sensor in your digital camera.

Coulomb:
The unit of electric charge, defined as the charge transported by a constant current of one ampere in one second.

Coulomb's law:
The magnitude of the force between two point charges is directly proportional to the product of their charges and inversely proportional to the square of the distance between their centers.

Current:
The flow of electric charge through a conductor, measured in amperes.

Current electricity:
The flow of electric charge through a conductor, typically in the form of an electric current.

DC (direct current):
Electric current that flows in only one direction, typically produced by batteries.

Decimal:
A numbering system based on base 10, using digits 0, 1, 2, 3, 4, 5, 6, 7, 8, and 9.

Desoldering braid:
A braided copper wire used to remove solder from electronic components and circuit boards.

Diode:
An electronic component that allows current to flow in one direction only.

Divider:
A circuit or component that divides a frequency by an integer. Dividing a wave in two produces a new wave with half the frequency of the original—the musical analog is rendering a note an octave lower.

Dogbotic:
The coolest audio studio in the world.

DPDT (double-pole double-throw):
A type of electrical switch that can control two circuits and switch between two positions.

DPST (double-pole single-throw):
A type of electrical switch that can control two circuits and switch between two positions, each connecting one circuit while disconnecting the other.

Drum machine:
An electronic musical instrument designed to imitate the sound of drums or percussion instruments.

Electric:
Relating to or using devices that either let electrons flow or don't.

Electricity:
The phenomenon when electric charge moves around.

Electron:
A subatomic particle with a negative electric charge, orbiting the nucleus of an atom.

Electronic:
Relating to or using devices that control the flow of electrons for various purposes, such as signal processing or power generation.

Envelope generator:
A circuit or module in a synthesizer that shapes an aspect of a sound over time (typically, the amplitude).

Faraday's law of induction:
A fundamental law of electromagnetism stating that a changing magnetic field induces an electromotive force (EMF) in a closed circuit.

Feedback:
The process of feeding a portion of the output of a system back into its input, often used to control or stabilize the system.

Filter:
A circuit or device that selectively allows certain frequencies to pass while attenuating others.

Flush cutters:
Precision cutting tools used to trim excess wire or component leads flush with the surface of a circuit board.

Frequency modulation:
The act of using the shape of one wave to change the frequency of another wave.

FM (frequency-modulation) synthesis:
A method of sound synthesis in which the frequency of one waveform is modulated by another at audio rate.

Frequency:
The number of occurrences of a repeating event per unit of time, typically measured in hertz (Hz).

Fundamental:
The lowest frequency of a harmonic series produced by a vibrating object, such as a musical instrument or a voice. It is the primary pitch perceived by the listener and forms the basis for the harmonic structure of a sound.

Gain:
The ratio between the input volume and the output volume in an amplifier circuit.

Gated oscillator:
An oscillator whose output is controlled by an external signal, allowing it to produce sound only when triggered.

Gator clip:
See "alligator clip."

Glide:
A parameter in synthesizers that controls the rate at which the pitch changes from one note to another.

Glissando:
A continuous slide upward or downward between two notes in music.

Ground:
A reference point in an electrical circuit, typically connected to the Earth or a common node.

Harmonic series:
The frequency components of a sound that are integer multiples of the fundamental frequency.

Harmony:
The combination of different musical pitches in order to create new, distinct musical ideas.

Heat-shrink:
A type of tubing that shrinks in diameter when heated, commonly used for insulating and protecting electrical connections.

Hertz (Hz):
The measurement of frequency, equal to one cycle per second. (A wave that oscillates 50 times per second has a frequency of 50 Hz.)

Inharmonic:
Not conforming to the harmonic series, thereby producing sounds with frequencies unrelated to the fundamental frequency.

Integrated circuit (IC):
A miniature electronic circuit consisting of semiconductor devices and passive components fabricated on a single piece of semiconductor material.

Interval:
The pitch difference between two notes.

Inverter:
A logic gate that outputs the opposite of its input.

Ion:
An atom or molecule with a net electric charge due to the loss or gain of one or more electrons.

Keyboard:
A set of keys or buttons used to input musical notes or data into electronic devices.

LED (light-emitting diode):
A semiconductor device that emits light when current flows through it.

Logic gate:
A basic building block of digital circuits that performs a logical operation on one or more binary inputs.

Low-frequency oscillator (LFO):
An oscillator that generates waveforms at frequencies below the audible range, commonly used to modulate other signals.

Luddite:
A person opposed to technological change or innovation, often specifically in reference to industrial machinery.

Magnetism:
The phenomenon by which materials exert attractive or repulsive forces on other materials.

Maintained switch:
A type of electrical switch that remains in its set position until physically actuated to change.

Mho:
An obsolete unit of electrical conductance, equivalent to the reciprocal of the ohm (Ω^{-1}).

Microphone:
A device used to convert sound waves into electrical signals.

Modulation:
The process of varying a carrier signal based on the characteristics of another signal.

Momentary switch:
A type of electrical switch that returns to its original position when released, typically used for temporary actions.

Mono (monophonic):
Referring to audio signals or systems that reproduce sound using a single channel.

Monophony:
Music or sound produced using a single note or pitch at a time.

NAND gate:
A logic gate that outputs false only when all its inputs are true.

Noise:
Unwanted sound or electrical interference that obscures desired signals. There is no physical distinction between desired sound and undesired noise. However, noise can have some helpful applications, including random number generation and creating a halfway decent snare drum sound (see the chapter Electronic Percussion).

Octave:
A musical interval spanning eight diatonic degrees, representing a doubling or halving of frequency.

Ohm (Ω):
The unit of electrical resistance, representing the resistance in a circuit where a potential difference of one volt produces a current of one ampere.

Ohm's law:
A fundamental principle in electrical engineering stating that the current through a conductor between two points is directly proportional to the voltage across the two points.

OR gate:
A logic gate that outputs true when at least one of its inputs is true.

Oscillator:
A circuit or device that generates an alternating voltage or current.

Overtone:
A higher frequency component of a complex sound, typically a whole-number multiple of the fundamental frequency.

Panning/panorama:
The distribution of a sound signal between the left and right audio channels in stereo systems.

Parallel connection:
A configuration of electrical components or devices in which the terminals of each component are connected to the same pair of nodes.

Partial:
Any of the individual frequency components that make up a complex tone. These partials include both the fundamental frequency and its harmonics or overtones.

Passive:
A circuit or component that does not require an external power source to operate.

PCB (printed circuit board):
A board made of insulating material with conductive pathways etched or printed on it, used to support and connect electronic components.

Phase:
The relationship in time between two or more waveforms, typically measured in degrees or radians.

Phase locking:
Synchronization of the phase of one oscillator to that of another oscillator or reference signal.

Photoresistor:
A type of resistor whose resistance varies with incident light intensity.

Piezo:
Short for *piezoelectric*, a term that refers to materials that generate an electric charge in response to mechanical stress, or vice versa.

Piezoelectricity:
The ability of certain materials to generate an electric charge in response to mechanical stress, or vice versa.

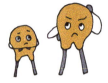

Pitch:
The perceived frequency of a sound, corresponding to its musical note or tone.

Polarized:
Having a preferred orientation or direction, typically used to describe components like capacitors or diodes.

Pole:
A point in an electrical circuit where multiple connections are made.

Polyphony:
Music or sound produced using multiple notes or pitches simultaneously.

Portamento:
A musical effect in which the pitch smoothly slides from one note to another.

Potentiometer (pot):
A variable resistor used to control electrical resistance and voltage levels.

Power amplifier:
An electronic device that increases the power of a signal to drive speakers or other loads.

Proton:
A subatomic particle with a positive electric charge, found in the nucleus of an atom.

Pulse:
1. Electronics: a brief burst of energy or signal, typically with a sharp rise and fall in amplitude.
2. Music: see "beat."

Pulse width:
The duration of a pulse signal, typically measured as a percentage of the total period.

RCA Cable:
A type of audio cable commonly used in consumer audio and video applications. It features a connector with a single pin for the signal wire and a surrounding metal shell serving as the

ground. RCA cables are typically used for unbalanced audio connections, such as connecting DVD players, gaming consoles, or amplifiers to speakers or TVs. They are not as resistant to interference as balanced cables and are best suited for short cable runs in home audio setups.

Relay:
An electromechanical switch operated by an electric current, typically used to control high-power circuits with low-power signals.

Resistance:
The opposition to the flow of electric current in a circuit, measured in ohms.

Resistor:
An electronic component that limits the flow of electric current in a circuit.

Resonance:
The reinforcement or amplification of a sound, vibration, or electric signal due to sympathetic vibration or oscillation.

Reverb (reverberation):
The persistence of sound in a particular space after the original sound source has stopped.

Rhythm:
The pattern of beats or pulses in music, often organized into regular groupings.

Ring modulation:
The act of multiplying one signal by the positive component of another signal, accomplished with a circuit that includes a ring of four diodes.

Sample and hold:
A circuit that samples an input signal at regular intervals and holds the value until the next sample is taken.

Sawtooth wave:
A waveform characterized by a linear rise in voltage followed by a rapid drop.

Schematic:
A diagrammatic representation of an electrical circuit or system, using standardized symbols to represent components and connections.

Sequencer:
A device or software program used to create and arrange musical sequences or patterns.

Series connection:
A configuration of electrical components or devices in which the output of one component is connected to the input of the next.

Sine wave:
A waveform characterized by a smooth, periodic oscillation, resembling the shape of the trigonometric sine function.

Skeuomorph:
A design element that retains ornamental elements of an original structure that were functionally necessary in the original.

Solder:
A low-melting-point metal alloy used to join metal surfaces.

Soldering iron/soldering pencil:
A tool used to heat solder and apply it to metal surfaces for soldering.

Sound:
Vibrations that travel through the air or another material and can be perceived by the ear.

SPDT (single-pole double-throw):
A type of electrical switch that can connect one input to either of two outputs.

Speaker:
A transducer that converts electrical signals into sound waves.

Spectrogram:
A visual representation of the spectrum of frequencies in a sound or other signal as they vary with time.

SPST (single-pole single-throw):
A type of electrical switch that can connect or disconnect a single input.

Square wave:
A waveform characterized by abrupt transitions between two levels, resembling a square shape.

Standoff:
A threaded support used to elevate and secure components above a circuit board.

Static electricity:
An imbalance of electric charges on the surface of an object, typically resulting from friction.

Stereo (stereophonic):
Referring to sound reproduction that uses two or more independent audio channels to create a sense of directionality or spatial imaging.

Stress (rhythm):
The emphasis or accent placed on certain beats or notes in music.

Subharmonic:
A frequency component of a complex sound that is an integer submultiple of the fundamental frequency.

Subtractive synthesis:
A method of sound synthesis in which complex sounds are generated by filtering and shaping harmonically rich waveforms.

Syncopation:
A rhythmic technique in which emphasis is placed on beats that are not typically accented.

Synthesizer:
An electronic musical instrument that generates and manipulates sound electronically.

Talkbox:
A device that allows a musician to shape sound by modulating the output of an instrument with their mouth.

Telharmonium:
An early electronic musical instrument invented by Thaddeus Cahill that was able to stream live music through telephone cables.

Theremin:
An electronic musical instrument controlled without physical contact by the position of the player's hands relative to two antennas.

Throw(s):
The number of positions a switch has.

Tie high or tie low:
The act of connecting an input of a digital circuit to a high voltage (logical 1) or low voltage (logical 0) to eliminate any ambiguous voltages between 0 and 1.

Tonewheel:
A rotating electromechanical device used in the Telharmonium and the Hammond organ to generate musical tones.

Transducer:
A device that converts one form of energy into another, such as electrical energy into mechanical motion (or vice versa) in a microphone, speaker, DC motor, piezo, or generator.

Transistor:
A semiconductor device with three or more terminals used for amplification, switching, or signal modulation.

Tremolo:
A modulation effect in music characterized by rapid fluctuations in volume or pitch.

Triangle wave:
A waveform characterized by linear transitions between rising and falling sawtooth shapes.

TRS (tip-ring-sleeve) cable:
A cable commonly used for stereo audio signals or balanced audio connections, featuring three conductors: a tip, a ring, and a sleeve.

Truth table:
A table that shows all possible input combinations and their corresponding outputs for a logic circuit.

TS (tip-sleeve) cable:
A cable commonly used for mono audio signals or unbalanced audio connections, featuring two conductors: a tip and a sleeve.

Unbalanced Audio:
A method of interconnecting audio equipment using cables that carry a single signal along

with a ground wire. Unlike balanced audio, unbalanced connections are more susceptible to interference and noise, especially over long cable runs. They are commonly found in consumer audio equipment and shorter cable runs where noise is less of a concern.

Undertone:
A frequency component of a complex sound that is lower in frequency than the fundamental frequency.

Vactrol:
A variable resistor using a light-dependent resistor (LDR) and a light source, commonly used for voltage control in electronic circuits.

Vacuum:
A space devoid of matter, typically created artificially for various purposes in electronic devices like vacuum tubes.

Vacuum tube:
An electronic device consisting of a sealed glass tube containing electrodes and a vacuum, used to amplify or switch electronic signals.

VCA (voltage-controlled amplifier):
An amplifier whose gain is controlled by an input voltage.

VCC:
The power-supply voltage used in digital integrated circuits.

VCF (voltage-controlled filter):
A filter whose frequency response is controlled by an input voltage.

VCO (voltage-controlled oscillator):
An oscillator whose frequency is controlled by an input voltage.

VDD:
The power-supply voltage used in digital integrated circuits.

Vibrato:
A musical effect produced by slight and rapid variations in pitch.

Voice:
A distinct melodic or harmonic part or section in music.

Volt (V):
The unit of electric potential or electromotive force, representing the potential difference between two points that will impart one joule of energy per coulomb of charge.

Voltage:
The electric potential difference between two points, typically measured in volts.

Voltage divider:
A circuit that divides a voltage into smaller voltages using resistors.

Voltage starve:
A technique in electronic music in which the voltage supply to a circuit is reduced, affecting the sound produced.

Waveform:
A graphical representation of the shape and amplitude of a wave, typically used to describe audio signals or electrical signals.

Wavetable synthesis:
A method of sound synthesis in which digital samples of waveforms are stored in a table and played back at various speeds to create different pitches and timbres.

Wire strippers:
Tools used to remove the insulation from the end of electrical wires, exposing the metal conductor for connection.

XLR Cable:
A type of audio cable commonly used for microphones and other balanced audio connections. It features a circular connector with three pins, each corresponding to a different signal wire: positive, negative (or inverse), and ground. XLR cables are known for their durability, secure connection, and ability to transmit balanced audio signals over long distances while minimizing interference.

XOR gate (exclusive OR gate):
A logic gate that outputs true only when exactly one of its inputs is true.

ABOUT THE AUTHORS

FUN FACT: YOU GOT RICKROLLED 27 TIMES BY THIS "NEVER GONNA GIVE YOU UP" OSCILLOGRAM. ;)

KIRK PEARSON is a composer and creative engineer based in the Bay Area. They spend a lot of time writing music for unconventional musical instruments and, every once in a while, inventing new ones. Kirk's sounds have appeared on prime-time television and in internationally touring installations, stage shows, animated cartoons, and hundreds of films, radio productions, and museum exhibits around the globe.

They are the founder and creative director of Dogbotic Labs, a Bay Area laboratory for inquiry-based art. Dogbotic Labs offers virtual workshops in a variety of oft-maligned creative topics, including e-textiles, DIY guitar pedals, film souping, datamoshing, animation, and creative coding. Through these workshops, Kirk has taught thousands of people from over fifty countries how to build a synthesizer.

Kirk's sounds have been featured in movies at the Sundance Film Festival, on animated shorts for Adult Swim, in PSAs for the UNHCR, and in installations at SXSW and the American Museum of Natural History. Several of their invented instruments can be heard in Rihanna's *Savage X Fenty Show Vol. 4*, onstage with dream pop band Vansire, and on permanent exhibit at the Oberlin Conservatory. In more trivial news, Kirk is a published paleontologist, an avid ice cream maker, and an enthusiastic juggler, and they once traveled 13,000 miles to film a tuna's journey from ocean to plate. Kirk lives in Berkeley with a loving (human) partner and their lizard, Andromeda. More at kirkpearson.com

SEAN RUSSELL HALLOWELL is a composer and video artist based in West Sonoma County, CA. He synthesizes experimental techniques developed from hand-built circuitry with a cosmic perspective on music's origins in number and time. His work has been featured at festivals and galleries across the US as well as in Mexico, Chile, Japan, South Korea, the UK, Belgium, Croatia, Poland, and Iceland. More at seanrussellhallowell.com

MAISY BYERLY (ILLUSTRATOR) originally hails from New Jersey, but now lives in Minneapolis with six hens, a rotating door of foster rabbits, and a 16-year-old cat. Her passion is people, places, and things with personality. She enjoys gathering people for printmaking days, turning corks into stamps and faces, and finding creative ways to encourage people to draw.

Since mid-2020, she has had the absolute pleasure of working with Dogbotic on designs for DIY synthesizers, cassette hacking workshops, and more. To Byerly, Dogbotic is a haven of lifelong learning and joyful exploration. Her first and only synthesizer, "Saltina," was exhibited at the Center for New Music in San Francisco. She has made around four zines and illustrated around four books, but still can't tell you what counts as a zine and when a book becomes a book. More at maisybyerly.com

SPECIAL THANKS

Tommy Marshall, for encouraging us to write a book in the first place, and for showing me what is possible with the DIY world.

Katy Luo, the first Dogbotic Labs manager, who believed in this project and gave it her all despite a devastating global pandemic.

Edward Shocker, director of Thingamajigs in Oakland, who has been nothing but a wonderful supportive force throughout this entire project.

Jeremiah Jenkins, for expertise on Han dynasty skeuomorphic pottery.

Kenneth John Taylor, the faceless author behind Ishkur's Guide to Electronic Music, an absolutely glorious and snarky Web 1.0 webpage that provided me with many hours of curious entertainment as a child. I'm also indebted to Ishkur for his argument about electronic music not having a birth date, which was such an interesting argument to my middle school self that I developed my whole philosophy of music around it.

Jasmine Bailey, our unflappably positive workshop coordinator, who never fails to lift our spirits during the hard times.

Zach Christy (zachchristy.com), who contributed the flipbook animation and has provided me with countless years of friendship and professional advice.

Roxanne Hoffman, the greatest business-development coordinator in the world, who is way funnier than she has any business being.

Maisy Byerly, who has created incredible art for us for years and for which we're forever indebted.

Jules Blom, for allowing us to use his beautiful illustrations of the Rotterdam metro.

Alexandra Kahn, for years of love and patience while listening to beeping sounds from the other room.

The following legends of homegrown synthesis, without whom I would never have come to sing the gospel of the humble CMOS chip: Joan Villaperros, José

Duarte, Sangbong Nam, Jin Sangtae, Nicolas Collins, Abby Aresty, Sudhu Tewari, Stanley Lunetta, Peter Blasser, Arthur Joly, Sebastian Tomczak, Elliot Williams, and Lisa Kori.

To the incredible teams at Make: and O'Reilly who have helped us through this process—especially editor Kevin Toyama, designer Juliann Brown, copy editor Mark Nichol, proofreader Carrie Bradley, and benevolent overlord Dale Dougherty.

To all the folks who teach or who have taught with Dogbotic Labs, we couldn't be prouder to work with you—Indranil Choudhury, Kittie Cooper, Callie Day, Nick Dunston, Sab Ghidossi, Aisha Loe, Alessandro Maione, Japhy Riddle, Chelsea Rowe, Alexander Taylor, and Linh My Truong.

To the composers, sound designers, musicians, engineers, and producers in the studio division of Dogbotic for the *buenas ondas*—Oreofe Aderibigbe, Hayden Arp, Casey Austin, Jonah Bobo, Anna Cataldo, Andrés Cervilla, Lydia Froncek, Vasundhara Gupta, Griffin Jennings, Charles Ryan, Kelsey Sharpe, and Kalia Vandever.

To all the students who've taken our workshops, believe in what we were doing, and have taken the time to tell us, thank you. The sheer amount of creativity and positive discourse that comes out of every workshop we teach is what gets us out of bed in the morning. We cannot thank you enough for giving us a newfound sense of purpose over these past few lonely years of COVID-19.

And finally, thank you to all the music/art/science teachers who, in some way or another, contributed to my general perspective on all the subjects in this book. In order of my education: Shlomo Pestcoe, Laurie Berkner, Meryl Danziger, Elvira Sullivan, Laurice Hwang, Joseph Meyers, Robert Apostle, Jim Pugliese, Omar Gonzalez, Heidi Reich, Irene Pease, Rob Owen, Karla Hubbard, Zeb Page (who I ripped off at least two jokes from in this book), Rebecca Leydon, Brian Alegant, Joo Won Park, Aurie Hsu, Zimoun, David Marín Romá, Gabriel Castillo Aguero, Bart Hopkin, and Garikayi Tirikoti. ♪ ♪ ♪

ABOUT DOGBOTIC

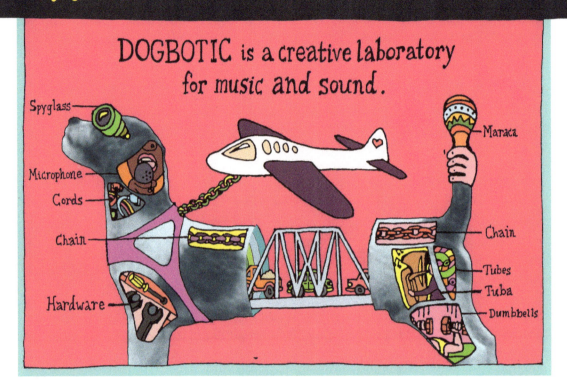

DOGBOTIC is a creative laboratory for music and sound.

- Spyglass
- Microphone
- Cords
- Chain
- Hardware
- Maraca
- Chain
- Tubes
- Tuba
- Dumbbells

Dogbotic is a Bay Area–based creative laboratory that makes weird and strange sounds for weird and strange people. We are a collective of musicians, technicians, composers, filmmakers, engineers, and visual artists who develop novel audio tools, invent our own musical instruments, and research new uses for emergent creative technology. Since 2020, we've offered educational workshops on oft-maligned creative topics, and we are dedicated to cultivating a friendly, supportive community where people from all walks of life can experiment and learn together.

A SELECTION OF OUR WORKSHOPS:

All of our workshops are virtual—we send the parts to you, and you learn with other students in small groups online. Most classes require no experience at all—just a sense of humor! Need-based financial aid is available for every virtual workshop we do.

CASSETTE HACKING: A Modern Musician's Guide to Mangling Magnetic Tape

A whimsical, socially distanced, and objectively magnetic workshop on making unconventional sounds with cassettes. Over the course of five weeks, we'll learn how to stitch together tape loops, circuit-bend a Walkman, control your cassettes with a modular synth, create a playable mini-Mellotron, and so much more! We'll dive into how magnetic tape actually works and how to exploit its quirks to make some original music, and we'll discuss the amazing cultural impact of the cassette tape, from ushering in an era of affordable home recording to instigating political revolutions. You'll learn a lot, make some cool gear, produce sounds that have definitely never existed before, and meet a bunch of friendly, creative folks. Cassette Hacking is an unconventional workshop for the musically curious, a class we really wish had existed back in 1995.

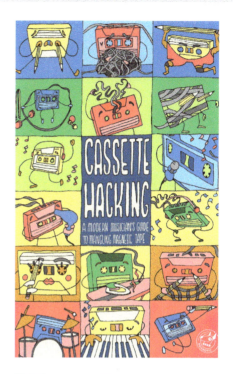

EAR RE-TRAINING: Media Manipulation for the Musical Mind

A music-composition workshop on media-bending experimental techniques. Join us for a 10-week hands-on odyssey through the landscape of music and media, equal parts technical and creative. Together, we'll craft a homemade radio to listen to overseas broadcasts, build a record player from scratch, conduct a cyanotypical seance, hack a greeting card into a functional digital sampler, recreate the reverb patterns of the world's greatest concert halls using a transducer, and so much more. Unlike any workshop we've assembled before, Ear Re-Training is a making-based class all about *why* music sounds the way it does, and how our obsession for audio preservation shapes our musical tastes. We earnestly believe this workshop will change the way you listen to sound.

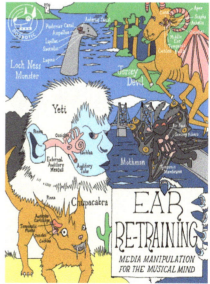

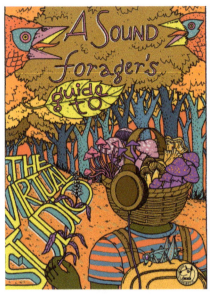

A SOUND FORAGER'S GUIDE TO THE VIRTUAL STUDIO

An experimental sound-design workshop to reacquaint yourself with your laptop and turn it into the last musical instrument you'll ever need. Over nine weeks, we'll explore the materiality of all the sounds we hear and make in the world as we capture, chop, stretch, knead, and otherwise transform audio to make everything from pop songs to strange soundscapes.

Stop using your laptops as glorified email-response machines once and for all. Using exclusively free software, we will unleash the powerful sound studio that's hidden inside every computer. Set foot inside your virtual studio to explore a world of weird and wonderful sounds.

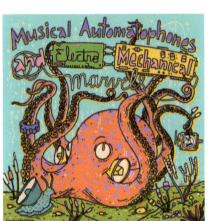

MUSICAL AUTOMATOPHONES (and Other Electro-Mechanical Marvels)

Craft a menagerie of self-actuating soundbots to tickle your senses, using homemade loudspeakers, motors, and all things electromagnetic. We will channel our inner Jedis and harness this force to shimmy, shake, and bounce our way to eclectic, one-of-a-kind rhythms, timbres, harmonies, and melodies. Using control mechanisms, sensors, and actuators, you will gain the technical skills and know-how to create your own automated music machines.

We will review a plethora of creative examples—both old and new—to inspire your inner inventor and help you carve out your own creative path. Above all else, we will experiment, play, and embrace happy accidents and wonky sounds. From lo-fi to high-tech, this class has something for anyone interested in venturing into the realm of music technology from the slightly offbeat angle of electromagnetic lutherie.

THREAD AND CIRCUITS: A Guide to Electro-Textiles

Needlepoint circuits. Knitted potentiometers. Wearable sensors and LEDs. Non-Euclidean crochet? A DIY beginners' virtual guide to the world of textile arts and measuring the world with microcontroller sensors, Thread and Circuits: A Guide to Electro-Textiles is designed to teach the basics of knitting, crochet, sewing, microcontroller interactivity, and Arduino code while also creating objects that allow these two mediums to interact regardless of your fiber, circuitry, or coding skills. (Watch out, Grandma and Silicon Valley startup tech bros!)

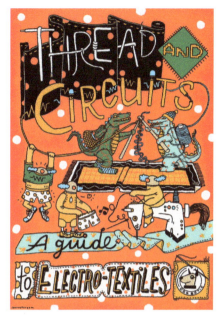

DIY SYNTHESIZERS FOR THE ELECTRONICALLY UNACQUAINTED

Learn about the art and culture of instrument building and the science of electricity through simple, fun, and intentionally ridiculous hands-on demonstrations. Our goal is to get you making strange and beautiful sounds as quickly as possible and to give you a career's worth of ideas on interesting ways to use them.

DIY RHYTHM WIDGETS FOR WONDERBOTS

We'll design our own analog drum circuits (everything from kicks to snares to hand claps), build a computer that can generate patterns on the fly, create an 808 (on a budget), return to a childlike state when we convert pots and pans into DIY drum triggers, and so, so, so much more.

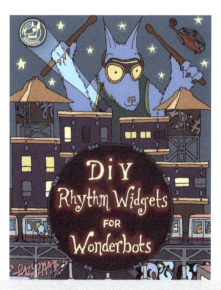

If you'd like to continue your journey with us, or are interested in collaborating with other people curious about electronic music, check out our virtual workshops at dogbotic.com.

www.ingramcontent.com/pod-product-compliance
Lightning Source LLC
Chambersburg PA
CBHW041007050326
40690CB00029B/5287